USING ALGEBRA

Ethan D. Bolker
UNIVERSITY OF MASSACHUSETTS, BOSTON

Wyndham Hall Press

USING ALGEBRA

by

Ethan Bolker
Professor of Mathematics
University of Massachusetts - Boston

Wyndham Hall Press printings

First printing, 1990
Second printing, 1991
Third printing, 1995

Library of Congress Catalog Card Number

90-050000

ISBN 1-55605-134-4 (paperback)

Printed in the United States of America

Wyndham Hall Press, Inc.
Bristol, IN 46507

Preface to the Preface

The times they are a changin'.
– Bob Dylan

Plus ça change, plus c'est la mème chose.

The first edition of *Using Algebra* [1] is now out of print. But the rationale for writing the book is as strong as it was a decade ago:

> "*The teaching of mathematics is shifting from preoccupation with inculcating routine skills to developing broad-based mathematical power.*
>
> "Mathematical power requires that students be able to discern relations, reason logically, and use a broad spectrum of mathematical methods to solve a wide variety of nonroutine problems. The repertoire of skills which now undergird mathematical power includes not only some tradition paper-and-pencil skills, but also many broader and more powerful capabilities. Today's student must be able to:
>
> - Perform mental calculations and estimates with proficiency;
> - Decide when an exact answer is needed and when an estimate is more appropriate;
> - Know which mathematical operations are appropriate in particular contexts;
> - Use a calculator correctly, confidently and appropriately;
> - Use tables, graphs, spreadsheets and statistical techniques to organize, interpret and present numerical information;
> - Judge the validity of quantitative results presented by others;

[1] Managed by three publishers: first Winthrop, absorbed by Little, Brown, then Scott Foresman.

- Use computer software for mathematical tasks;
- Formulate specific questions from vague problems;
- Select effective problem-solving strategies." [2]

This book addresses most of these points in some way. So I'm pleased that Wyndham Hall is reprinting it.

Some small corrections have been made to the original, but I haven't tried to bring it all up to date. Calculators are cheaper now. (Is the projection in Problem 7 on page 122 correct?) The problems here on inflation deal in the double digits appropriate when I wrote the first draft. There's nothing on global warming, or the AIDS epidemic. I leave it to you to build problems in the style of mine on these and other new topics. All you need do is read the newspaper every day.

In this new format, it will be easier to keep up to date. If you send me problems you construct I'll include some in future Wyndham Hall editions.

Ethan Bolker
December, 1989

[2] *Everybody Counts: A Report to the Nation on the Future of Mathematics Education*, Board on Mathematical Sciences, Mathematical Sciences Education Board, National Research Council, National Academy Press, Washington D.C., 1989, pages 82-83.

Preface

> The justification for a concept lies exclusively in its clear and unambiguous relation to facts that can be experienced.
>
> – Albert Einstein

Useful, straightforward, thought-provoking, and fun: qualities college students rarely ascribe to mathematics. But this text presents just such mathematics, for students who have studied some algebra and want to know more, either to pursue one of the social or life sciences or to satisfy their pure intellectual curiosity.

The book consists entirely of problems in which mathematics models familiar phenomena, like automobile fuel consumption, depreciation, population growth, inflation, and the spread of epidemics. To investigate such problems requires algebra: a thorough study of linear equations, unit calculations, the estimation of large numbers. logarithms, e, and the trigonometry of right triangles. Throughout the book the nature of the problems studied dictates the approach to the mathematics–setting up models and drawing inferences, but little formal manipulation and no proofs identified as such. I believe that the abstract truths that make mathematics beautiful and useful are best left implicit at this level. I have found that students tend to agree.

To master real data, the book teaches the use of the scientific calculator in parallel with the mathematics. Features are introduced as their need is felt. The calculator builds confidence because it frees the user to concentrate on mathematics without worrying about arithmetic. Students learn a new skill while enjoying concrete approaches to some traditionally troublesome abstractions like inverse functions and the rounding of answers to the appropriate number of significant digits.

On pace and coverage: When I teach real problems at a reasonable rate students find the mathematics clearly helpful in solving those problems–they learn more, remember more, and enjoy more than they do when I hurry. If I try to cover too much material I revert to teaching algorithms rather than how to think about problems. Then student interest flags and less is learned more painfully. My right pace for a one-semester course allows me to cover all of Parts I and II, on Linear and Exponential Models (treating Chapters 6, 7, 14, and 15 somewhat skimpily), and either Part III (Quadratic Models) or Part IV (Trigonometry).

In the five years since I first taught a course based on these ideas I have learned from both students and colleagues exposed to this growing text. Particular thanks go to Peter Ash, Bernice Auslander, Don Babcock, Valerie Collins, Dan Comenetz, Pamela Dodds, Marilyn Frankenstein, Matt Gaffney, Paul Mason, Michael Monette, Betty O'Neil, Earl Porter, Barbara Ruth, Geza Schay, Robert Seeley, Mary Shaner, Lorraine Turner, Tan VoVan, and Wing Aero. Students who suggested problems may find those problems here, with their names attached. Andrew Gleason added a thoughtful reading of an early draft of this material to years of talk about mathematics and its philosophy, pedagogy, and exposition. Alex Greene's editorial enthusiasm was just the spur I needed to turn rough notes and ambitious ideas into this finished book. Bob Morris kept UNIX running and answered formatting questions while Benjamin Bolker typed the manuscript. Little, Brown's reviewers, Dan Kemp (South Dakota State University), E. James Peake (Iowa State University of Science and Technology), Shirley Sorensen (University of Maryland), and, particularly, Charles Jepsen (Grinnell College), were receptive and helpful. Ian Irvine and Dave Lynch at Little, Brown managed the editing and production with open-minded vigor. My wife Joan was there through some hard times—I'm glad we can share this good one. My children, Jessica and Benjamin, were there, too. Growing and learning and laughing with them have shaped me and hence this book. They are part of why I enjoyed writing it so much, so it's for them.

To the Student

The bashful cannot learn, nor the irritable teach.

– Hillel, *Sayings of the Fathers*, II, 6

With common sense and some algebra you can understand the world better than you can with common sense alone. That is the theme of this book on using algebra. Almost every problem is an application, a description in words of a piece of the world. No problems appear just in order to illustrate mathematical ideas. We will study the mathematics of automobile fuel economy, of inflation, of radioactive decay, and of population growth. You will learn to estimate numbers that are hard to measure, and to manipulate very large and very small numbers.

To study this material you must already know some algebra, and you will master more while you learn to see the world in a mathematical way. But you will probably find less emphasis on the rules of algebra than you are used to. We concentrate more on how to read problems and make sense of what they say than on the laws that tell how to move mathematical symbols around on paper. The variables will rarely be the traditional x and y. They will more often be P (for, say, electric power consumption, measured in kWh per person per year) or T (for time, measured in years or seconds).

The text is a sequence of worked examples designed to introduce new ideas in concrete settings. Read them. Listen to lectures about them. Then work problems. Mull them over, play with them, and talk about them with your friends. Just watching is not enough. Mathematics is not for spectators. If something in the text puzzles you, ask about it. Perhaps my estimate or my models do not match your experience. Perhaps something could be more clearly explained. Question. Argue. Look for understanding, not for revealed truth. Each problem, each technique, each answer is supposed to make sense. You may find that idea strange. Many students have been convinced by their experiences that the last thing mathematics can be is sensible. If you're one of them, I hope I can change your mind.

While working through this book you will learn to use a calculator.

For many of you that will be a new tool, and novelty makes learning exciting. We can use real data rather than simple numbers in the problems. Moreover, using a calculator does more than make arithmetic tolerable. It provides new routes toward understanding difficult ideas.

> The material that directly addresses your increasingly sophisticated use of the calculator is in indentations like this one.
>
> You will need a scientific or slide rule calculator, one with a key labeled "*y* to the power *x*," or something similar. Right now (1982) such calculators cost between $15 and $40. Your calculator must provide
>
> both common and natural logarithms (LOG and LN),
> parentheses (or RPN, as on Hewlett-Packard calculators),
> scientific notation, to display large numbers,
> at least one memory location.
>
> If it uses rechargeable (or 1000-hour) batteries you will not be replacing them frequently. You will find more than one memory register very useful, though not essential. Other features may meet your long-term needs. Some calculators have built-in statistical tools. Some do common business calculations efficiently. Some are particularly suited for scientific work. The most exciting option you could select would be programmability, although a programmable calculator might cost more than $40 and is not necessary for this material. Buy the best calculator you can afford. Plan to keep it.

This book was designed to cover useful and interesting material, worth studying for its own sake. It may not be suitable preparation for more advanced work in mathematics. In particular, this is not a standard precalculus text, although it covers many of the topics included in such texts. It can, though, prepare you to take a short, applied calculus course. I hope that you enjoy learning the mathematics in this book. I have tried to find examples to interest you if you care about biology, economics, management, or sociology, or if you want to know how mathematics can help you see the world in a new way.

Annotated Contents

Part I

Linear Models

CHAPTER 1

Introduction: The Wing Aero Corporation

And the crooked shall be made straight, and the rough places plain.

– *Isaiah*, XL, 4

Understanding when two variables are linearly related
The significance of the slope and the intercept, and their units
The first use of your calculator / Graphs / Solving linear equations / Finding data outside this book and the classroom

EXAMPLE 1. WING AERO

The Wing Aero company makes solar energy collector panels. The company is small, and the number of panels they make each month varies. The cost of materials (copper and Plexiglas) is \$375 per panel. Moreover, no matter how many panels are ordered, rent and salaries must be paid. Each month those fixed costs, or overhead, amount to \$5000. Discuss how the total monthly cost depends on the number of panels produced that month.

We start by concentrating on the materials cost, and temporarily ignore the overhead. If, for example, Wing Aero makes 63 panels in September their materials cost is

$$375 \frac{\$}{\text{panel}} * 63 \text{ panels} = 23\ 625\$. \tag{1.1}$$

Equation 1.1 may look odd: it is full of words as well as numbers. The words are called *units*; they are included to make the meaning clear. We rarely deal with pure numbers. In Equation 1.1 we did more than multiply 375 by 63. The 63 means 63 *panels*. The 23 625 means 23 625 *dollars*; for a while we will write that as 23 625\$, the way it is read aloud, rather than as \$23 625. The units for the number 375 are subtler; they are *dollars per panel*. Since the word *per* always means *divide by*, *dollars per panel* becomes \$/*panel*. In Equation 1.1 we multiplied 375 *dollars per panel* by 63 *panels*. The result was 23 625 *dollars*. The word *panel(s)* cancels, leaving just \$. The cancelation of units is analogous to the canceling you do in numerical and literal fractions.

If Wing Aero made 12 panels in August, the materials that month would have cost

$$375 \frac{\$}{\cancel{\text{panel}}} * 12 \cancel{\text{panels}} = 4500\$. \qquad (1.2)$$

We can summarize these calculations algebraically by introducing variables. Let N be the number of panels manufactured in a month. The *units* of N are *panels*. Then the materials cost in that month is

$$\text{materials cost} = 375N,$$

or, rewritten with units,

$$\text{materials cost } (\$) = 375 \left(\frac{\$}{\cancel{\text{panel}}} \right) * N(\cancel{\text{panels}}).$$

However, Wing Aero must pay overhead in addition to the cost of materials. Thus the total cost when 63 panels are made in a month is

$$\begin{aligned} \text{overhead} + \text{materials cost} &= 5\,000\$ + 23\,625\$ \\ &= 28\,625\$. \end{aligned}$$

When only 12 panels are made the total cost is

$$5000\$ + 4500\$ = 9500\$.$$

Let's name the total cost C; its units are \$. Then C, the cost of doing business for a month, *depends on* N, the number of panels produced in that month. The dependence can be written in words as

$$\text{total cost} = \text{overhead} + 375 \frac{\text{dollars}}{\text{panel}} * \#\text{ of panels}. \qquad (1.3)$$

Algebraically, this becomes

$$C = 5000 + 375N, \tag{1.4}$$

or, with units displayed,

$$C(\$) = 5000(\$) + 375\left(\frac{\$}{\text{panel}}\right) * N(\text{panels}). \tag{1.5}$$

We can use Equation 1.4 to prepare Table 1.1, showing how C changes with N. Since we are thinking of how C depends on N, C is the *dependent variable*, and N, whose variation causes C to change, is the *independent variable.* Because we use the value of N to calculate the value of C, tradition requires that in Table 1.1 N be on the left and C on the right.

The entries for C opposite $N = 0$ and $N = 47$ are new. Look now at the last line of Table 1.1. What happens to poor Wing Aero in a month when no panels are ordered? Then $N = 0$ in Equation 1.4. The cost C is 5000\$, which is the overhead alone.

To find the entry in the table for $N = 47$, calculate

$$\text{cost for 47 panels} = 5000 + 375 * 47. \tag{1.6}$$

> Does Equation 1.6 say "Add 5000 to 375, then multiply the result by 47"? That gives 252 625. Or does it mean "Add 5000 to the result of multiplying 375 by 47"? The answer then is 22 625. We know that the second answer is correct. Why?
>
> First, the meaning of Equation 1.6 comes from Equation 1.3 or 1.5. There the units tell you to do the multiplication first, then the addition.
>
> Second, in this problem *common sense* alone tells you that 252 625\$ is much too large a number for the cost when 47 panels are made in a month. After all, a 63-panel month costs Wing Aero only about 29 000\$. You should always check to see that answers *make sense*.

Table 1.1 Wing Aero's monthly costs

N (number of panels made)	C (total monthly cost in \$)
12	9 500
47	22 625
63	28 625
0	5 000

Third, the first people who did a lot of arithmetic found it useful to adopt the convention that in equations like 1.4 the multiplications are always done first. When the addition was meant to come first they used parentheses. We observe this convention.

Try out your calculator now. Be sure you can make it give the correct answer in Equation 1.6. Some calculators know the convention that multiplication takes precedence over addition, some do not. See if yours does by punching in

5000 [+] 375 [*] 47 [=].

(If your calculator has no [=] key it is probably from Hewlett-Packard, uses a logical structure known as RPN, and needs no [=] key.) If you get the wrong answer, experiment with

5000 [+] [(] 375 [*] 47 [)] [=],

and

375 [*] 47 [+] 5000 [=].

We can now describe how C depends on N in two ways: with the formula in Equation 1.4 or with Table 1.1. There is a third, even more useful way. It is to draw the graph of the dependency as shown in Figure 1.1. Here the *horizontal axis* is for N, the *independent variable*, and the *vertical axis* is for C, the *dependent variable*.

The striking fact about the five data points in Figure 1.1 is that they seem to lie on a straight line. In fact, they not only *seem* to, they do. That is why we say that *C depends linearly on N*. To describe the line that exhibits the dependence you need only two pieces of information: the overhead, measured in \$, and the materials cost, measured in \$/panel. The overhead is the *C intercept*, the value of C when $N = 0$ panels. It is the height at which the graph crosses the vertical or C axis; it is a constant that does not change when N changes. It is the first term on the right of the equals sign in Equation 1.4; its units are \$, the units of C.

The other number we need, the materials cost of 375 \$/panel, is the coefficient of N in Equation 1.4. Its units are (units of C)/(units of N). That coefficient is called the *slope* of the line.

The situation we have just encountered is very general. A quantity P *depends linearly* on another quantity Q whenever there are constants I and m such that

$$P = I + mQ. \tag{1.7}$$

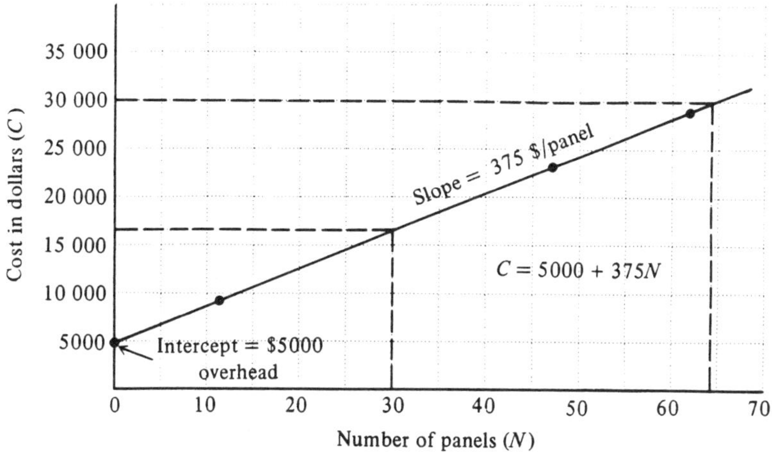

Figure 1.1 Wing Aero monthly costs

Then the graph of the dependence will be a straight line. I is the P intercept (the value of P when $Q = 0$); its units are the units of P. The constant m is the *slope*; its units are (units of P)/(units of Q). (Clearly, I is an abbreviation for *intercept*. I do not know why slopes are often called m, but they are.) The slope measures the rate at which P increases when Q increases.

The straight line in Figure 1.1 allows us, or Wing Aero's president, to answer (approximately) some cost questions without using algebra, or even arithmetic. If we wish to know approximately what costs will be in a month when 30 panels are built we find $N = 30$ on the horizontal axis, travel vertically (on the dotted line in the figure) to the graph, and then move left to the vertical C axis. There we read

$$C \sim 16\,000\$.$$

> The symbol $\sim$ means *is approximately equal to*. Often an approximate answer is all that is needed in a real problem. Then you don't need a calculator. But even when an exact answer is required, it is wise to find an approximate one first. If your guess is close to the exact answer, you can be pretty sure you made no mistake. If it is not, either you misunderstood the problem and guessed poorly, or you miscalculated. Rework the problem until your guess and the exact answer more or less agree.

The exact cost when $N = 30$ panels can be found using Equation 1.4; it is 16 250\$. The guess from the graph was quite close, and easier to find.

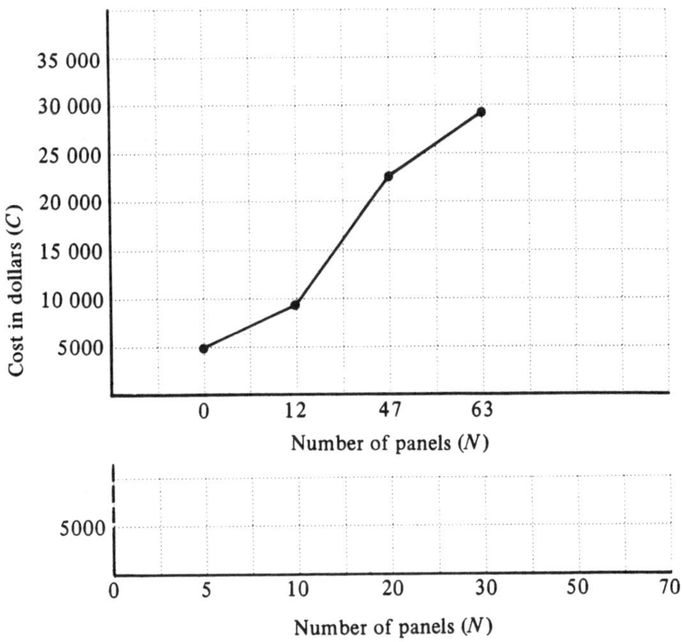

Figure 1.2 Dishonest horizontal scales

You will be drawing graphs often as you learn applied algebra. Here is some advice.

1. Use graph paper.
2. Label the axes and state the units in which the variables are measured. Choose the vertical and horizontal scales carefully. The data should fit nicely on the page. You may never have had to think about choosing scales in earlier algebra courses, where, usually, each box on the graph paper counts for one unit.
3. The scales must be honest, like rulers. Put 0 in its proper place, and space the numbers evenly. (Figure 1.2 shows two botched scales for the N axis in this example.)

EXAMPLE 2. LOOKING BACKWARD

Because old customers have not paid their bills, Wing Aero knows they have only 30 000$ to spend in November. How many panels can they manufacture?

This problem is new to us because we are given C and asked to find N. We begin solving it by using the graph to find an approximate answer.

Starting at $C = 30\,000\$$ on the vertical axis in Figure 1.1, we move right to the line and then down, where we see that the value of N is approximately 65 panels.

To find N exactly, we substitute $C = 30\,000\$$ in Equation 1.4. This gives

$$30\,000 = 5\,000 + 375N, \tag{1.8}$$

which we must solve for N. You learned how to do that in your first algebra course. Subtract 5000 from each side, giving

$$30\,000 - 5\,000 = 375N.$$

Exchange the two sides, since we are more comfortable with the unknown N on the left:

$$375N = 30\,000 - 5\,000. \tag{1.9}$$

Then divide each side by 375, so that

$$N = \frac{30\,000 - 5\,000}{375} = 66.67.$$

To understand the answer better, repeat the algebra with units included. Equation 1.8 becomes

$$30\,000\$ = 5\,000\$ + 375\,\frac{\$}{\text{panel}} * N\text{ panels}$$

Then Equation 1.9 tells us

$$375\,\frac{\$}{\text{panel}} * N\text{ panels} = 25\,000\$. \tag{1.10}$$

This means that after the overhead has been paid, Wing Aero will have 25 000$ to spend on materials. Dividing each side of Equation 1.10 by 375 $/panel tells us that

$$N\text{ panels} = \frac{25\,000\$}{375\,\dfrac{\$}{\text{panel}}} = 66.67\text{ panels}. \tag{1.11}$$

In Equation 1.11 we again manipulate the words expressing units as if they were numbers, using the rules you have learned for fractions. To divide by a fraction, you invert and then multiply:

$$\frac{\$}{\dfrac{\$}{\text{panel}}} = \cancel{\$} * \frac{\text{panel}}{\cancel{\$}} = \text{panels.}$$

Thus for 30 000$ Wing Aero can manufacture 66.67 panels in that month. Not quite. The answer our algebra produced makes no sense in the problem. Wing Aero can't make a fraction of a panel, and so we must round our answer. The whole number nearest to 66.67 is 67. But that's not the right answer, either. By rounding up from 66.67 to 67 we increase the cost to more than the 30 000$ allowed. To fit the constraints of the problem we must round down to 66 panels. If Wing Aero has 30 000$ to spend in a given month they can manufacture only 66 panels. (Even that answer would be correct only if Wing Aero bought material for one panel at a time.)

In many problems in this book you will have to solve equations like Equation 1.8. But you will also learn that solving equations is only about one third of the work called for in solving problems. The *first* third is setting up the equations to be solved–deciding how to translate a real situation, like the relation between Wing Aero's cost and production figures, into algebra. The *second* third is solving the equations. The *last* third, without which the first two thirds are of little use, is to translate the numerical answer into a statement about the original situation and to see that it makes sense. Thus the solution to Example 2 is not the number 66.67; it is the sentence "If Wing Aero has 30 000$ to spend in a month they can make 66 panels."

EXAMPLE 3. LOOKING BACKWARD SYSTEMATICALLY

Study how the number of panels Wing Aero can make in a month depends on how much they can spend for materials and overhead.

We have already answered this question when $C = 30\,000\$$ by solving Equation 1.8 for N. To study how N depends on C more systematically we must solve Equation 1.4 for *N in terms of C*. Start by subtracting 5000 from each side, giving

$$375N = C - 5000. \tag{1.12}$$

The right side of Equation 1.12 tells us how much money Wing Aero can spend on materials after the overhead is paid. Then dividing each side by 375 yields

$$N = \frac{C - 5000}{375}. \tag{1.13}$$

That answers the question asked: we have discovered how N depends on C. We can reproduce the answer we found in Example 2 by substituting $C = 30\,000$ in Equation 1.13. The units in that equation are correct too. The units of $C - 5000$ are \$ − \$, or \$, and the units of the 375 are \$/panel; so the fraction on the right has the units *panels*, as in Equation 1.11.

We shall finish this example by rewriting Equation 1.13 in several ways.

First, dividing by 375 is the same as multiplying by 1/375, which is approximately 0.00267, so that

$$N = 0.00267 * (C - 5000). \tag{1.14}$$

> To find 1/375 you can enter
>
> 1 [÷] 375
>
> on your calculator. But there is a faster way. Most scientific calculators have a *reciprocal* or [1/x] key, with which you can find 1/375 directly:
>
> 375 [1/x].
>
> The display will show
>
> 0.0026667.
>
> The number of 6's will vary from machine to machine. Notice the 0 *preceding* the decimal point. It appears on many calculators; it makes the decimal point easier to find. Develop the habit of writing it down.

Now, multiplying through the parentheses in Equation 1.14, and noticing that $0.00267 * 5000 = 13.33$, we get

$$\begin{aligned} N &= 0.00267C - 13.33 \\ &= -13.33 + 0.00267C. \end{aligned} \tag{1.15}$$

The second version of Equation 1.15 has the same form as Equation 1.7. It shows that N depends linearly on C. Figure 1.3 displays the

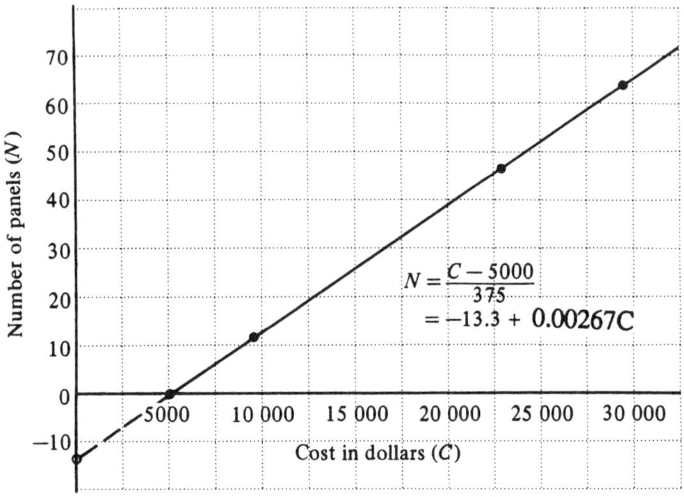

Figure 1.3 Wing Aero monthly costs

graph of that dependence. The intercept is the value of N when $C = 0$. It is, with its units, the negative number

$$-13.33 \text{ panels.}$$

What does that mean? Setting $C = 0$ makes sense in Equation 1.15, but it makes no sense in the problem. That is why we draw the part of the graph to the left of $C = 5000$ as a dotted line. One way to interpret the negative intercept is to observe that it represents the 13.33 panels Wing Aero could afford to buy material for if they did not have to pay overhead.

The slope of the line in Figure 1.3 can be read from Equation 1.15. It is

$$0.00267 \ \frac{\text{panels}}{\$}.$$

Its units, which are (units of N)/(units of C), tell you its meaning. Each extra dollar Wing Aero spends buys the material for 0.00267 panels. That's a small fraction of a panel, because the materials cost, 375 \$/panel, is a number much larger than 1.

EXAMPLE 4. DEPRECIATION

Wing Aero owns a panel stamping machine that cost them 26 000\$. It depreciates (for tax purposes) at the rate of 1800\$ per year. What is it worth after 5 years and 3 months? When is it worthless?

To answer the questions asked, we first decide what the variables are, and name them. In this example, the value of the panel stamper depends on how old it is. We shall use V for the value (its units are \$) and A for its age (measured in years). Then V depends on A. The statement of the problem tells us that the dependence is linear. The clues are the words *rate* and *per*. Thus we seek an equation of the form

$$V = \text{intercept} + \text{slope} * A \tag{1.16}$$

with which to calculate V when A is known.

The slope is the rate given:

$$-1800 \frac{\$}{\text{year}}.$$

Its units are (units of V)/(units of A), as expected. It is negative, because as A increases V decreases. The older the panel stamper the less it is worth.

The intercept is the value of V when $A = 0$. That we recognize as the purchase price of 26 000\$. It is what the machine is worth before it has begun to age. Thus, to find V when A is given use the equation

$$V = 26\,000 - 1800A. \tag{1.17}$$

Figure 1.4 shows the graph of this dependence. The line slants down to the right, exhibiting the negative slope.

Now, having understood the problem, we can answer the particular questions asked. At age 5 years and 3 months, $A = 5.25$ years. (Three months is 1/4 or 0.25 of a year, not 0.3 year!) The graph in Figure 1.4

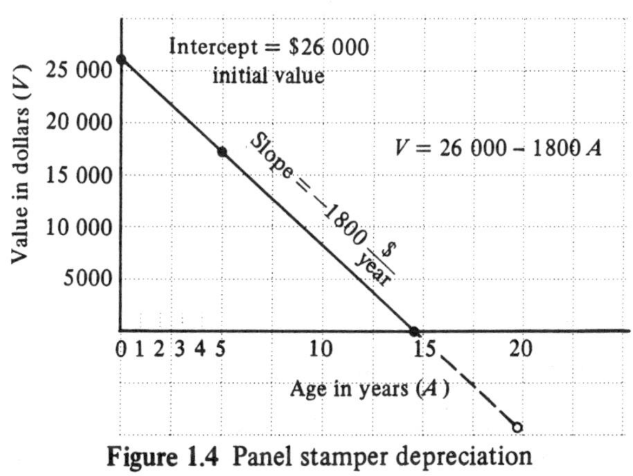

Figure 1.4 Panel stamper depreciation

gives an estimate of between 16 000$ and 17 000$ for V. To find V more precisely, substitute $A = 5.25$ in Equation 1.17 to find

$$V = 26\,000 - 1800 * 5.25 = 16\,550\$.$$

The graph shows that $V = 0$ when A is between 14 and 15 years. That is the age at which the machine is worthless. (Worthless, that is, for tax purposes; it may still be working well.) To find that age exactly, substitute $V = 0$ in Equation 1.17 and solve for A:

$$0 = 26\,000 - 1800A$$

so

$$1800A = 26\,000$$

and

$$A = \frac{26\,000}{1800} = 14.44 \text{ years.}$$

This is about 14 years, 5 months, since

$$0.44 \text{ year} = 0.44 \text{ year} * 12 \frac{\text{months}}{\text{year}} \sim 5 \text{ months.}$$

If we were to ask about the value of the stamper when it was 20 years old, Equation 1.17, which makes sense when $A = 20$, would tell us that

$$V = 26\,000 - 1800 * 20 = -10\,000\$.$$

This makes little sense. That point is shown on the dotted part of the line in Figure 1.4. The entire line is described by Equation 1.17, but only the solid part says anything meaningful about Wing Aero's panel stamper.

PROBLEMS

Read all the problems, even those not assigned.
Answers to starred problems are in the appendix.

1. Find Wing Aero's cost in a 60-panel month, first approximately, using the graph, and then exactly. Now 60 is just twice 30, but

the cost in a 60-panel month is *not* twice the \$16 250 cost in a 30-panel month. Why not? Compare the materials costs alone for the 30- and 60-panel months.

2. If Wing Aero can spend \$19K in a month, how many panels can they produce? First guess from the graph in Figure 1.1 or Figure 1.3, then compute. (The letter K is an abbreviation for *thousand*, so \$19K means \$19 000.)

*3. A vacation in England costs \$720 for round-trip air fare and \$85 for each day you stay there. Name the appropriate variables, write the linear relationship, identify the slope and the intercept (with their units), and draw a graph. How much will a two-week vacation cost? If you have \$2100 to spend, how long a vacation in England can you take?

4. The price of sugar changed rapidly from July 1, 1974 to December 31, 1974 after a poor harvest. The price P (in dollars) of a 5-pound bag depended very closely on the time t (in months after July 1, 1974) according to the equation

$$P = 0.80 + 0.45t.$$

Draw a graph of this dependence. Find P on July 1, 1974 and on September 15, 1974. When was $P = \$3.00$? Identify the slope and the intercept, with their units.

*5. A monthly telephone bill is computed at \$16.50 plus 7 cents per message unit.

(a) Name the appropriate variables and write the linear relationship between them. (Do not confuse the letter you choose to stand for the number of message units with the units themselves.)

(b) Identify the slope and intercept with their units. Draw a graph.

(c) What will the bill be in a month in which 238 message units were used?

(d) How many message units were used in a month in which the bill was \$28.33?

(e) Solve the equation you found in (a) for the number of message units in terms of the monthly bill.

6. When my brother-in-law uses his darkroom it takes him 25 minutes to set up his chemicals and clean up when he's done. In between he needs 1.5 minutes per negative.

(a) Study how the total time he spends depends on the number of negatives he prints. Choose names for your variables and tell what units they are measured in. Write an equation. Draw a graph, properly labeled. Identify the slope and intercept, with their units. What total time will he need to print 43 negatives?

(b) Now solve the equation you have found in order to show how the number of negatives printed depends on the time spent. Draw the graph of that relationship; identify the slope and intercept with their units. If he has two hours to spend, how many negatives can he print?

7. Nicolette Brown jogs every day. She warms up for 5 minutes, and then runs smoothly at the rate of 8 minutes/mile.

(a) Write an equation and draw a graph showing how the total time she spends depends on the distance she runs. How long will it take her to warm up and run 6.2 miles?

(b) Now rework your equation so that it shows how the distance she runs depends on the time she spends exercising. Draw a new graph. What is the slope of the line? What is Ms. Brown's speed in miles/minute? How far can she run if she has half an hour to spend?

*8. A new refrigerator sells for $475. Running it costs 8 cents per day. Write an equation showing how the total cost of refrigeration depends on the length of time the refrigerator has been in use. How much does it cost to keep your food cold for the first year? When will you have paid as much in daily costs as it cost you to buy the refrigerator?

9. A car costs $7200 when new. Suppose its value V, in dollars, depends linearly on its age T, in years, and decreases at the rate of 680 $/year. Write an equation showing how V depends on T. Identify the slope and the intercept, with their units. Draw a graph. How much is the car worth when it is 6.5 years old? When is it worth half its original value? When is it worthless? Use your equation to find the value of the car after 20 years. Comment on the meaning of your answer.

10. When you travel a steady 55 miles per hour the distance you cover depends on the time you travel. Name the variables and their units and write the equation connecting them. The dependence is linear. Why is the intercept 0 miles? What is the slope? (This problem anticipates Example 1 of Chapter 4.)

11. Massachusetts levies a 5 percent sales tax. That is, you pay 5% or 0.05 *dollar of tax* per *dollar of purchase*. Find a formula showing how to compute the tax when you know the amount of purchase.

Notice that the relationship is linear with intercept 0. (This problem anticipates Example 3 of Chapter 4.)

*12. Union dues are set at \$85 plus half of one percent of wages. Find a formula showing how dues depend on wages. Identify the units of all the quantities. What are the dues to be paid on wages of \$5000 and \$10 000? Draw a graph. If Mr. VoVan paid dues of \$213 how much did he earn?

13. In Massachusetts the income tax you pay is (approximately) 5 percent of the amount of your income in excess of \$1000. (In fact, the deduction is not the same \$1000 for all taxpayers; the 5% rate applies only to income earned in the state, and a surcharge is added after the tax is computed. We shall ignore these complications in this problem.) Find a formula showing how to compute the tax in terms of the income. Graph the result. Be sure your graph shows the correct tax for incomes less than \$1000. Find the tax on an income of \$30 000. If Ms. Diamond paid \$900 in taxes, what was her income?

14. Several years ago it was fashionable to think not only that we should tax the wealthy more than the poor, but should help out the poor, too. Poverty for a family of four was defined as an annual income of \$4000. We shall study a highly simplified version of what was proposed. In our version, if you earn more than \$4000, you pay 17% of the amount over \$4000. If you earn less than \$4000, the government pays you 17% of the amount by which you fall short. If you earn \$4000 you are thought to be just well enough off and neither pay any tax nor receive any subsidy. Let T represent the tax paid (in dollars) on an income of I dollars.

(a) What financial transactions take place when income is \$10 000? \$3000? \$4000? \$0?

(b) Explain why the formula

$$T = 0.17(I - 4000)$$

gives the right answer in all cases.

(c) Draw a graph of the linear dependence of T on I exhibited in (b). Find the slope and intercept, with their units, and interpret them.

(d) Explain why the rule: "Tax is 17% of the difference between I and \$4000" gives the same results as: "Collect 17% of I in taxes, and then have the government pay each wage earner \$680."

15. The Trezh Car Rental Company charges an initial \$32, and then 27 cents (or \$0.27) per mile. Write an equation showing how the total rental cost depends on the number of miles driven. Do the same for the Siva company's costs. They rent a similar car for \$28 plus 30 cents per mile. Compute the costs for each rental for trips of 50, 100, and 200 miles. Draw the two graphs on the same set of coordinate axes. Discuss the relative merits of renting from Trezh or Siva. For what length of trip are the costs equal? (This problem anticipates Example 1 of Chapter 6.)

*16. On a winter afternoon in Boston the temperature was $-4°$C (Celsius) at 5 o'clock and was dropping at a rate of 1.5° per hour. Write an equation showing the dependence of the temperature on the time. (Measure the time in hours *since 5 o'clock.*) Draw a graph. What was the temperature at 6:15? At 4:00? When was it 0°C? When $-10°$C?

17. Guy Calvert owns a small fishing boat. To fish for a day it costs him \$90 in wages, diesel oil, and other incidental expenses. Any fish he catches he can sell for \$1.30 per pound. Draw a graph and write an equation showing how his profit depends on the size of his catch. How much must he catch to break even? Show that point on your graph. What is his profit on a day when he catches 193 pounds? 265 pounds? What is his profit *per pound* on each of those days? What happens to his profit per pound as fishing gets better and better?

18. It costs \$2 000 000 (2 million dollars) to design a calculator and set up an assembly line and \$12 to manufacture each one. Write the linear equation that tells how the total production cost depends on the number of calculators manufactured. Why must many be sold before a profit can be made?

*19. The graph in Figure 1.5 shows how the sales of the Red Gear Co. grew during the last decade. Estimate the intercept and the slope, with their units, and write a formula. Use the graph and your formula to find sales in 1968, 1970, 1975, and (predicted) 1990. (This problem anticipates Example 4 of Chapter 2.)

20. Table 1.2 shows the population of the United States in millions of people from 1800 to 1960.

Find the population in 1970 and in 1980 in a reference library. Draw a graph displaying the growth of the United States population during the period from 1800 to the present. Which of Figures 1.6a or 1.6b is a wiser choice for the horizontal scale? Is the population growth linear?

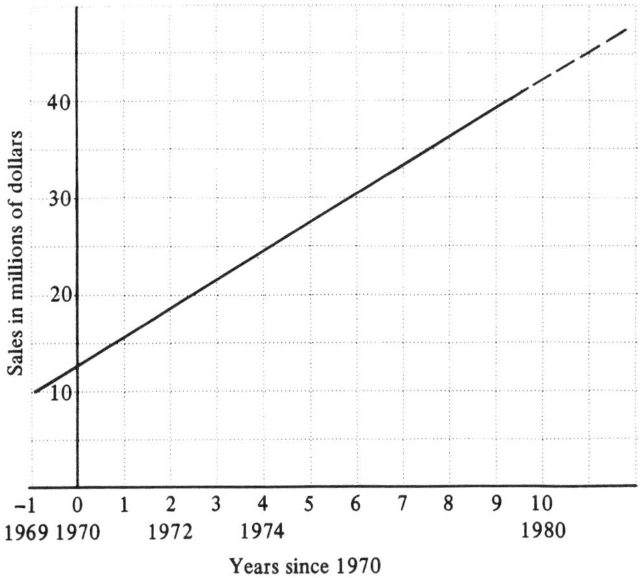

Figure 1.5 Red Gear Co. Sales

Table 1.2 The population of the United States

Year	Population in millions
1800	5.3
1810	7.2
1820	9.6
1830	12.9
1840	17.1
1850	23.2
1860	31.4
1870	39.8
1880	50.2
1890	62.9
1900	76.0
1910	92.0
1920	105.7
1930	122.8
1940	131.7
1950	150.7
1960	170.3

Source: Encyclopaedia Britannica, 1966

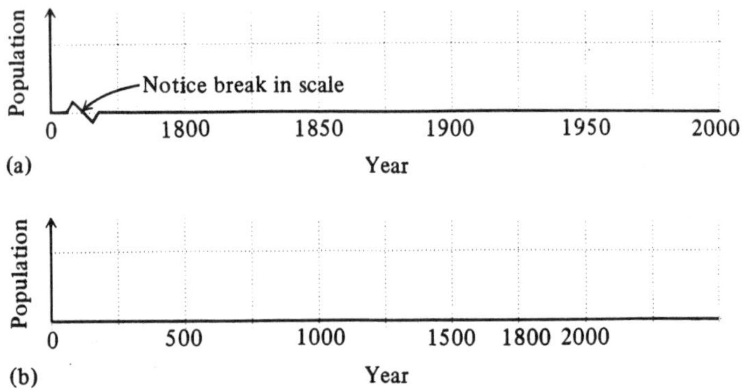

Figure 1.6 Two possible scales for graphing United States population

21. Criticize the graph in Figure 1.7, which is intended to display the linear relationship between the columns in Table 1.1 on page 5.

22. Excerpt from *Science*, May 22, 1981, p. 901: "[Data from Hiroshima] . . . reveal a clear relationship between doses of radiation received and ill effects. That relationship is defined by a linear equation: An increase in dose above the natural background radiation correlates with a proportional increase in ill effects."

    What are the variables implicitly defined in this paragraph? What is the intercept? The slope?

23. Find two graphs, one linear and one not, in recent newspapers or periodicals. For each, identify the units of the quantities being discussed.

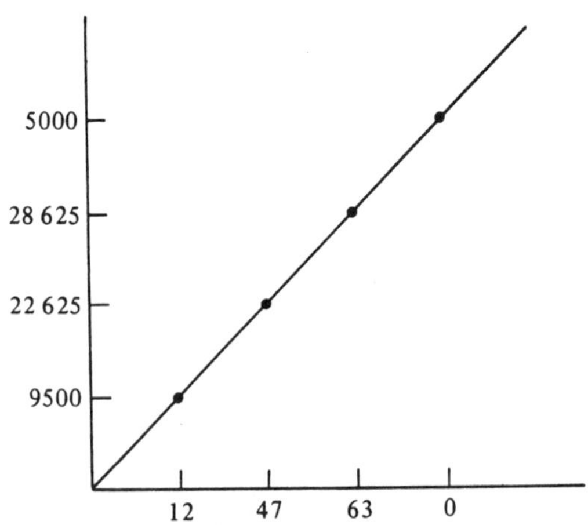

Figure 1.7 A graph to be criticized

24. Suppose the equation

$$J = \frac{0.7 - T}{5.8}$$

describes the dependence of J on T. Rewrite the equation so that it shows explicitly that the dependence is linear, and find the slope and intercept. Then solve either version of the equation for T in terms of J, and find the slope and intercept for that linear dependence.

*25. Solve each of the following equations for R in terms of T:

(a) $T = 4.6 + 0.17R$ (b) $T = -4.6 + 0.17R$

(c) $T = 4.6 - 0.17R$ (d) $T = -4.6 - 0.17R$

(e) $T = I + 0.17R$ (f) $T = 4.6 + mR$

(g) $T = I + mR$

26. Solve each of these equations for W in terms of Y, and for Y in terms of W:

(a) $Y + W = 0$ (b) $2Y + 3W = 0$

(c) $Y - W = 0$ (d) $2Y - 3W = 0$

(e) $Y + W + 1 = 0$ (f) $2Y - 3W + 1 = 0$

(g) $aY + bW = 0$ (h) $aY + bW + c = 0$

27. Solve for P in terms of Q: $Q = 4.6 + 0.17(P - 7.5)$.

*28. Solve for Z in terms of W: $W = 4.6 + \dfrac{Z - 7.5}{1.3}$.

29. Solve for M in terms of L: $L = \dfrac{4.6 + M - 7.5}{1.3}$.

30. (a) Solve $C = 5000 + mN$ for N in terms of C and m. Compare your answer with Equation 1.13.

(b) Solve $C = F + 375N$ for N in terms of C and F. Compare your answer with Equation 1.13.

(c) Solve $C = F + mN$ for N in terms of C, m, and F. Again compare with Equation 1.13.

*31. Solve for M in terms of m: $m = \dfrac{3 - M}{5 - M}$.

32. Suppose that $A = mP + M(1 - P)$.

(a) Solve for m in terms of A, M, and P.

(b) Solve for M in terms of A, m, and P.

(c) Solve for P in terms of A, m, and M.

33. Estimate each of the following numbers; then use your calculator to find the exact answer. When parentheses are called for, try both with and without them.

(a) 12 + 34 * 56 (b) (12 + 34) * 56 (c) 34 * 56 + 12
(d) (34 * 56) + 12 (e) 12 * 34 + 56 (f) 12 * (34 + 56)
(g) 12 − 34 * 56 (h) (12 − 34) * 56 (i) 12 − 34 − 56
(j) 12 − (34 − 56) (k) 12 * 34 * 56 (l) (12 − 34) − 56

CHAPTER 2

The Line Through Two Points

> And five of you shall chase a hundred, and a hundred of you shall chase ten thousand; and your enemies shall fall before you.
>
> – *Leviticus* XXVI, 8

Recovering the slope and intercept of a line from numerical data or a graph / Using the calculator effectively and efficiently

EXAMPLE 1. BERKELEY PLASTICS

The Wing Aero Corporation (WA) learns that a competitor, Berkeley Plastics (BP), manufactured 12 panels in July for a total cost of \$8044 and 46 panels in August for \$22 052. Wing Aero wants to know Berkeley Plastics' monthly overhead and their materials cost. They have hired us as their mathematical consultant in this game of corporate espionage.

First we choose the variables. Let C (units: dollars) be the total cost in a month in which BP makes N panels. Since for Wing Aero C depends linearly on N we make the same assumption about BP; so

$$C = \text{intercept} + \text{slope} * N. \tag{2.1}$$

Our job is to find the intercept, which represents the monthly overhead, and the slope, which is the materials cost in \$/panel. We are given the following data:

N (No. of panels)	C (Total cost, \$)
12	8 044
46	22 052

Observe that in August $46 - 12 = 34$ *extra* panels were produced, compared with July. The *extra* cost for producing those panels was \$22 052 − \$8044 = \$14 008. That extra cost was all for materials, none for overhead, so the materials for 34 panels cost \$14 008. Hence the materials cost is

$$\frac{\text{extra cost}}{\text{extra panels}} = \frac{\$14\,008}{34 \text{ panels}} = 412 \frac{\$}{\text{panel}}.$$

Before we find BP's overhead, the intercept, let us reflect on what we've discovered so far: BP's materials cost is greater than WA's 375 \$/panel. A 12-panel month would cost WA $5000 + 12 * 375 =$ \$9500, which is greater than BP's cost of \$8400, in spite of BP's higher materials cost. We can thus predict that BP's overhead must be lower than WA's \$5000. Just how much is that overhead?

If we substitute the now known slope of 412 \$/panel in Equation 2.1 we get

$$C = \text{intercept} + 412N. \tag{2.2}$$

Subtracting $412N$ from each side gives

$$\text{intercept (or overhead)} = C - 412N. \tag{2.3}$$

We know that $C =$ \$8044 when $N = 12$ panels. Substituting in Equation 2.3 tells us that

$$\text{intercept} = \$8044 - \$412 * 12 = \$3100.$$

Thus we will tell Wing Aero that Berkeley Plastics' monthly overhead is \$3100, their materials cost is 412 \$/panel, and the linear equation with which to calculate their total monthly cost is

$$C = 3100 + 412N.$$

To check, substitute $N = 46$ panels and see whether the known value $C =$ \$22 052 results. It does. Figure 2.1 is a graph showing BP's and Wing Aero's costs.

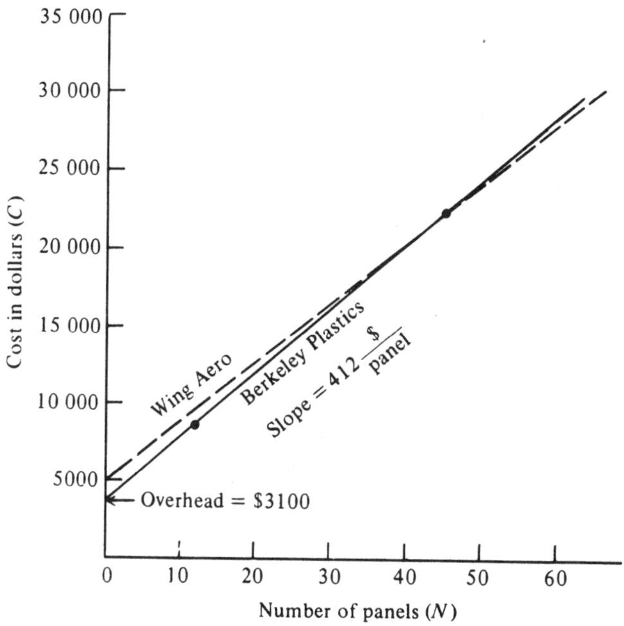

Figure 2.1 Berkeley Plastics' monthly cost compared to that of Wing Aero

EXAMPLE 2. AIRPLANE FUEL CONSUMPTION

A jet airplane uses a fixed quantity of kerosene for a takeoff and a landing (combined) and, when in the air, a certain quantity per mile. If a 400-mile trip requires 1455 gallons and a 260-mile trip uses 1077 gallons, find the fuel used for a takeoff and landing, and the rate of fuel consumption during flight. How much kerosene will a 1230-mile trip require?

The first sentence of this problem tells us that the kerosene consumption F (measured in gallons) depends linearly on T, the length of trip (measured in miles). Thus

$$F = \text{intercept} + \text{slope} * T. \tag{2.4}$$

The intercept is the amount of fuel used just to take off and land without going anywhere: the value of F when $T = 0$. The slope describes the fuel use in flight; its units will be gallons/mile. We will postpone the question about the 1230-mile flight until we have found the slope and intercept.

We first apply the method developed in Example 1 to find the slope in gallons per mile. It is given by the ratio (extra kerosene)/(extra mileage). Therefore

$$\begin{aligned} \text{slope} &= \frac{(1455-1077)\text{ gallons}}{(400-260)\text{ miles}} \\ &= \frac{(1455-1077)}{(400-260)}\ \frac{\text{gal}}{\text{mi}} \qquad (2.5)\\ &= 2.7\text{ gal/mi.} \end{aligned}$$

Do all the calculator arithmetic at one time. There is no need to compute or write or remember the ingredients 1455 − 1077 or 400 − 260 separately. Experiment to see which parentheses in this calculation you really need.

To find the intercept, substitute the slope 2.7 gal/mi in Equation 2.4 and rearrange to get

$$\text{intercept} = F - 2.7T. \qquad (2.6)$$

Substituting $F = 1455$ when $T = 400$ in this equation tells us that

$$\text{intercept} = 1455 - 2.7 * 400 = 375 \text{ gal.}$$

This calculation too should be made without writing down any intermediate results. Your calculator may give the right answer if you simply enter

1455 [−] 2.7 [*] 400 [=].

If it won't, either use parentheses:

1455 [−] [(] 2.7 [*] 400 [)] [=],

or do the multiplication first:

2.7 [*] 400 [CHS] [+] 1455 [=].

The key [CHS] changes the sign of the number displayed. Some calculators label it [+/−]. You cannot use the subtraction key [−] for this purpose.

If you did not turn your calculator off after checking Equation 2.5, then it still displayed the slope 2.7 when you reached

Equation 2.6. You did not have to reenter it. If you have a memory register in your calculator you should save the slope in each of these problems. If you have more than one memory register, save the intercept too. (It is not too hard to reenter two-digit numbers, but later, when the numbers are longer and rounding error becomes significant, you will want to have developed the habit of using the memory registers.)

The final fuel-consumption equation is thus

$$F = 375 + 2.7T. \tag{2.7}$$

It takes 375 gallons of kerosene to take off and land, and 2.7 gal/mi to fly. A 1230-mile trip will require 375 + 2.7 * 1230 = 3696 gal, or nearly 3700 gallons of fuel.

EXAMPLE 3. LINEAR DEPRECIATION

A commercial refrigerator costs $23 000 new and will be worth $500 as scrap in ten and a half years. Discuss how its value depends on its age.

The variables in this problem are V, the value of the refrigerator in dollars, which depends on Y, its age in years. It is common to assume, for tax purposes, that V depends linearly on Y. The graph is shown in Figure 2.2. The equation has the form

$$V = \text{intercept} + \text{slope} * Y. \tag{2.8}$$

We are given

$$V\,(\text{when } Y = 0) = 23\,000 \tag{2.9}$$

and

$$V\,(\text{when } Y = 10.5) = 500. \tag{2.10}$$

The expression V (when Y = 10.5) in Equation 2.10 is often abbreviated as $V(10.5)$ and read "V of 10.5." The parentheses *do not* indicate multiplication. We shall use this abbreviation from time to time so that you will become accustomed to it.

What are the slope and intercept? In each of the first two problems we found the slope first, and then the intercept. However, our work here is easier if we notice that we already know the intercept. It is $V(0)$,

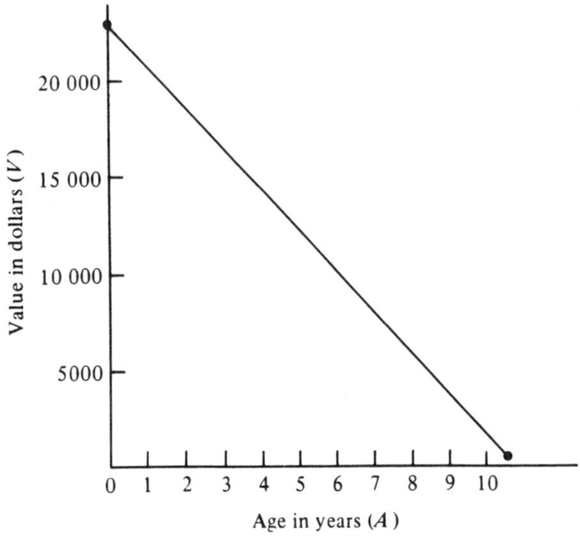

Figure 2.2 Refrigerator depreciation

the value when $Y = 0$, which we are told is \$23 000. We find the slope in the usual way.

$$\begin{aligned} \text{slope} &= \frac{\text{change in value}}{\text{change in age}} \\ &= \frac{V(10.5) - V(0)}{10.5 - 0} = \frac{500 - 23\ 000}{10.5} \\ &= -2142.8571 \ \$/\text{yr} \\ &= -2140 \ \$/\text{yr (for practical purposes).} \end{aligned}$$

The units of the slope are \$/yr. The slope is *negative* since the value is decreasing. The calculation in the denominator was easy since the change in age from 0 to 10.5 years is just 10.5 years. Then we rounded the value of the slope to the nearest ten dollars. Thus the value of the refrigerator in terms of its age is given by the equation

$$V = 23\ 000 - 2140Y.$$

EXAMPLE 4. RED GEAR CO.

Figure 1.5 on page 19 shows how the sales of the Red Gear Co. grew in the 11-year period from 1968 to 1979. Since the graph is a straight line we can find an equation for it of the form

$$\text{SALES} = \text{intercept} + \text{slope } Y \tag{2.11}$$

where the SALES are measured in millions of dollars and the time Y in years *since* 1970. Since $Y = 0$ in 1970, the intercept is the SALES figure for 1970, which we see on the graph is about 14 million dollars. To find the slope we must find a second point on the graph, in order to compute the change in the SALES. We can use 1979. That is, $Y = 9$; the sales then were 40 million dollars. Thus

$$\begin{aligned}
\text{slope} &= \frac{\text{change in SALES}}{\text{change in } Y} \\
&= \frac{\text{1979 sales} - \text{1970 sales}}{9 - 0} \\
&= \frac{(40 - 14) \text{ million \$}}{9 \text{ years}} \\
&= 2.9 \, \frac{\text{million \$}}{\text{year}}.
\end{aligned}$$

Hence the equation for the graph in Figure 1.5 is

$$\text{SALES} = 14 + 2.9Y.$$

PROBLEMS

1. It costs \$2.75 to send a 4-pound package from Boston to Milwaukee, and \$5.15 to send a 10-pound package.
   - **(a)** Assume the shipping cost depends linearly on the weight. Find a formula exhibiting that dependence. Identify the variables, the slope and the intercept, and their units.
   - **(b)** Use your answer to (a) to find out what it would cost to send a 5.5-pound package.
   - **(c)** Solve the equation obtained in (a) to give a formula for the weight of a package in terms of the cost to send it to Milwaukee. How heavy a package could you send for \$7.00?

*2. Write each of the following expressions on one line, using the slash (/) for division and parentheses when necessary. Then calculate the value of each:

(a) $\dfrac{12.81 - 4.67}{2.59 - 0.63}$ **(b)** $\dfrac{12.81}{2.59} - \dfrac{4.67}{0.63}$

(c) $\dfrac{12.81 - 4.67}{2.59} - 0.63$ **(d)** $12.81 - \dfrac{4.67}{2.59 - 0.63}$

(e) What do you think is the point of this exercise?

*3. The cost of heating a house in Boston for the winter depends linearly on the number of *degree days*. When the number of degree days is 4000 the total cost is \$1050. When there are 5600 degree days the cost is \$1425. Name the variables, find a linear equation relating them, and draw a graph of the relationship. What are the units of the slope and intercept? What would it cost to heat the house in a winter in which there were 5300 degree days? Optional questions for those who live in a cold climate: What is a degree day? How many of them were there last winter?

4. To eat at the Faculty Club one pays a fixed, one-time membership fee, and then a constant amount per lunch. If 30 lunches cost a total of \$112.50, while 47 lunches cost \$159.25, find the membership fee and the cost per lunch. Write a formula for computing the cost C of L lunches. Then solve for L in terms of C. Find C when $L = 35$. If Professor Schay's lunch budget for the year is \$325, how often can he eat at the Faculty Club instead of carrying his lunch from home?

*5. The tax value V (in dollars) of a bulldozer declines linearly from \$35 000 when it is new to \$7000 when it is 7 years old. Find an equation showing how V depends on the age of the bulldozer. Draw its graph. When is the tax value \$10 000?

6. The amount of gas energy (measured in *therms*) used for cooking in a month is equal to a fixed amount for the pilot light plus some for each hour the burners are in use. In a month when the burners are on for 51 hours, 4.403 therms are consumed. When the burners are used 63 hours in a month, 5.219 therms are used. Find a formula expressing the number of therms used as a function of burner use. Draw its graph. Optional question: Look up the definition of the unit *therm* in which natural gas energy is measured.

*7. Wing Aero manufactures "Softer Saucers" in addition to solar panels. When Ben's friend Will works for Wing Aero he finds he can make 78 Softer Saucers in an hour and 148 in an hour and 40 minutes. How fast can he make saucers (in minutes per saucer) once he has set up the machine? How long does it take him to set up? What is his work rate in saucers per minute?

8. Suppose it costs 9 cents per mile for repairs and maintenance (gasoline excluded) to drive a 4-year-old car, but only 3 cents per mile for a new one. Assume that the maintenance cost depends linearly on the age of the car.

(a) Write a formula for finding the maintenance cost given the age. Identify the slope and the intercept and their units.

(b) How much will maintenance cost for an 8-year-old car?

(c) Solve the equation obtained in (a) to give a formula for calculating the age of a car given the cost of maintaining it. Use the equation to discover how old a car is if 7 cents per mile is needed to maintain it.

*9. When I woke up at 7:15 A.M. there were 1.4 inches of snow on the ground. At noon the depth was 4.3 inches. Assuming linear dependence, write a formula showing how the depth of snow depends on time. At what rate was the snow falling? What was the depth at 10 A.M.? When did the snowstorm start?

10. On the Fahrenheit temperature scale water freezes at 32° and boils at 212°. (Those peculiar numbers came about because the designer wanted 0° to represent the freezing point of salt water–which it does–and 100° normal human body temperature. There he missed a little: 98.6° is correct.) On the Celsius scale (formerly called centigrade) water freezes at 0° and boils at 100°. The relation between the scales is linear.

(a) From the information given find a formula for converting °C to °F and one for converting °F to °C. Graph each.

(b) What is normal human body temperature in °C?

(c) At what Celsius temperature does salt water freeze?

(d) At what temperature will a Celsius thermometer and a Fahrenheit thermometer agree?

11. The profit a movie theater makes in a week depends on the number of customers. In a week when there are 1372 customers the profit is \$1790; 1115 customers yield a profit of \$1310.

(a) Name the variables and find a linear equation relating them.

(b) Draw a graph. Identify the units and explain the meaning of the slope and the intercept.

(c) Find the profit in a good week–one with 2000 customers.

(d) What is the break-even point? That is, how many customers does the theater need to begin making a profit?

(e) Find a formula expressing the number of customers in terms of the profit.

*12. (a) A vacuum cleaner salesperson makes a small weekly salary and a large commission. When he sells 13 vacuum cleaners in a week he takes home \$182.65 (before taxes); when he sells 20 he makes \$244.95. Find his salary and commission. Sketch a graph and write its equation.

(b) A second salesperson is on straight commission (no fixed initial salary). When she sells 13 vacuum cleaners she too makes \$182.65. How much will she make when she sells 20? (This problem anticipates Example 2 of Chapter 4.)

13. Suppose the crime rate (in incidents per 10 000 people) depends linearly on the population density (in thousands of people per square mile). In one city the population density is 5.1 thousand people/square mile and the crime rate is 3.2 incidents per ten thousand, and in another city the corresponding figures are 8.7 and 4.6. Write a formula showing how the crime rate depends on population density. Identify the units of all quantities expressed.

*14. In a session at the computer word processor I can type 4 pages in 50 minutes and 9 pages in an hour and a half. Find my typing rate in minutes per page and the time, independent of the number of pages, which it takes me to log on and get ready for work. How long will it take me to type 20 pages? How much can I do in two hours? In two days?

15. Suppose that in the verse from *Leviticus* that opens this chapter, the relation between numbers of chasers and numbers of chased is linear.

(a) Write an equation relating these two variables. How many could 120 chase? How many chasers would be needed to confront an enemy numbering 7000?

(b) What is the intercept in the equation you have found? Does it make sense? How many could 3 chase?

(c) Show that the equation

$$\text{No. of chased} = 100 + (\text{slope you found}) * (\text{No. of chasers} - 5)$$

gives the answers to the questions in (a).
(This problem anticipates Example 1 of Chapter 3.)

*16. Suppose the airplane in Example 2 has tanks that hold 4290 gallons. How long a trip can it make? How much fuel would be required for a 2500-mile trip (with a refueling stop)?

17. Total solar panel sales for the year in Berkeley will be 1850 panels. Wing Aero and Berkeley Plastics are the only suppliers.

(a) Write an equation showing how Wing Aero's yearly sales depend on those of Berkeley Plastics.

(b) Graph the dependence. Identify and interpret the slope and intercept of the line.

*18. Suppose that a quantity Q depends linearly on another quantity P. Find the equation exhibiting this dependence if:

(a) $Q = 8$ when $P = 0$, and $P = 5$ when $Q = 0$. Draw the graph.

(b) $Q = a$ when $P = 0$, and $P = b$ when $Q = 0$.

19. It costs Leigh Rhodes \$11.30 to take a 7-mile cab ride and \$18.30 to go 12 miles.

(a) Find the linear equation governing cab fares. Interpret the intercept and the slope.

(b) Find the cost for trips of 9.2 miles and 15.3 miles.

(c) How far can she travel for \$10.00?

(d) Since taxi meters compute fares in multiples of \$0.10, rather than in pennies, your answers to (b) and (c) are probably a little off. If the rate is quoted as "10 cents for each * * * of a mile," what fraction does * * * stand for? What are the correct answers to (b) and (c)?

20. Sam's racketball club costs \$12 less to join but \$1.25 more per hour of play than Susan's. For how many hours' play is Sam's cheaper?

21. The firm of White and Daughters knows that the *marginal cost* of manufacturing widgets is \$0.53. That means that once they have set up shop for the week their production cost is \$0.53/widget.

(a) If the total cost in a week in which they manufacture 10 000 widgets is \$8100, find their weekly overhead (the cost of setting up and running the shop for that week).

(b) What is a widget? Use a good dictionary to find out.

22. Use a calculator to learn some algebra in the following examples:

(a) Calculate: $(1/2.7) + (1/5.6)$.

(b) Calculate: $1/(2.7 + 5.6)$.

(c) Your answers in (a) and (b) show that, in general,

$$\frac{1}{x} + \frac{1}{y} \text{ is } \textit{not} \text{ the same as } \frac{1}{x + y}.$$

Make up several more examples that demonstrate the same point.

(d) Does it ever seem to be true that $(1/x) + (1/y)$ *does* equal

$1/(x+y)$? Suppose x and y are both positive. Which expression is always larger than $(1/x)$? What about the other one?

23. A common error in algebra is to cancel the A in the fraction

$$\frac{Ax}{A+y}$$

giving either

$$\frac{x}{1+y} \quad \text{or} \quad \frac{x}{y}.$$

(a) Substitute some numbers for A, x, and y, and use your calculator to show that neither of the two supposed answers is correct.

(b) Use the same method to see what happens if you try to cancel the A in the expression

$$\frac{A+x}{A+y}.$$

Of course you *can* cancel the A in $\frac{Ax}{Ay}$.

*24. Suppose a quantity Q depends linearly on another quantity, P. Find an equation giving Q in terms of P, and another giving P in terms of Q, given the following values for P and Q:

(a) $Q = 7$ when $P = 9$ $\quad$ $Q = 22$ when $P = 14$

(b) $P = 7$ when $Q = 9$ $\quad$ $P = 22$ when $Q = 14$

(c) $Q = 7.3$ when $P = 9.8$ $\quad$ $Q = 22.7$ when $P = 14.9$

(d) $Q = A$ when $P = 9$ $\quad$ $Q = A'$ when $P = 14$

(e) $Q = 7$ when $P = B$ $\quad$ $Q = 22$ when $P = B'$

(f) $Q = A$ when $P = B$ $\quad$ $Q = A'$ when $P = B'$

25. (a) If the slope of a line [measured in (units of W)/(units of Z)] is 0.056, and $W = 0.031$ when $Z = 0.086$, find an equation with which to calculate W when Z is given.

(b) Solve the same problem when 0.056 is replaced by the constant M.

(c) Solve the same problem when, in addition, the constants B and A replace 0.031 and 0.086 respectively.

CHAPTER 3

Interpolation and Extrapolation

> There is something fascinating about science. One gets such wholesale returns of conjecture out of such a trifling investment of fact.
>
> – Mark Twain, *Life on the Mississippi*

How to study a line without using its intercept
Reading between the lines of a table / Rounding
Learning to question the validity of a linear model

EXAMPLE 1. DIESEL WATER PUMPS

A water pump with a 7.5-horsepower diesel motor can move 18.5 gallons of water per minute and one with a 17-hp motor can pump 44 gal/min. In the range from 5 to 30 hp the pumping rate depends linearly on the power of the motor. Find the rate for pumps powered by a 10-hp and by a 25-hp motor. How powerful a motor would be needed to pump 30 gal/min?

Let R represent the pumping rate in gal/min, and P the power of the motor in hp. We know that R depends linearly on P, and that $R(7.5) = 18.5$, while $R(17) = 44$. (Recall that "$R(17) = 44$" means "the value of R when $P = 17$ hp is 44 gal/min.") The slope of the line that graphs the dependence of R on P is found by calculating

$$\begin{aligned}\frac{\text{change in } R}{\text{change in } P} &= \frac{R(17) - R(7.5)}{17 - 7.5} \\ &= \frac{44 - 18.5}{17 - 7.5} \qquad (3.1) \\ &= 2.6842 \ \frac{\text{gal/min}}{\text{hp}}\end{aligned}$$

Thus each increase in motor power of 1 hp causes the pumping rate to increase by approximately 2.7 gal/min.

> Sometimes calculators are too precise. The answer to the numerical problem posed in Equation 3.1 is indeed 2.6482 . . . , but it is nonsense to report that the slope of the line we are studying is 2.6842 (gal/min)/hp. The only way we could arrive at such an accurate answer would be to begin with data that accurate. In this example we are entitled to report only that the slope is 2.7 (gal/min)/hp.
>
> The general rule is: *Report answers* only as accurate as your data.

A rough way to describe the accuracy of a number is to count *significant digits*. Those are the figures other than the zeros used to locate the decimal point. In this example the numbers 44, 17, and 7.5 each have two significant digits; 18.5 has three. The calculator shows the slope as 2.6842 . . . , a number with at least 5 significant digits. But one cannot get more accuracy out of a calculation than one puts in. That is why we round the slope to a number with two significant digits—2.7.

> The numbers 30 268 000, 3.0268, and 0.0030268 each have 5 significant digits. Rounding each to 4 significant digits gives 30 270 000, 3.027, and 0.003027.
>
> Some calculators can be instructed to display numbers rounded to a prescribed number of significant digits while maintaining more accurate values internally for further computations.

We could now find the intercept for the line in Figure 3.1, which exhibits the dependence of R on P. That figure shows that the pumping rate when $P = 0$ hp is approximately -2 gal/min. This is a ridiculous statement about water pumps even though the graph is correct. We shall solve the problem without mentioning the intercept by working entirely in the range 5–30 hp, within which the dependence is linear. The method we are about to illustrate is called *interpolation*, from the Latin roots for *between* and *to polish*.

To find $R(10)$, the pumping rate for a 10-hp motor, use Equation 3.1 again in a new way. We know that

$$\frac{\text{change in } R}{\text{change in } P} = 2.6842\ \frac{\text{gal/min}}{\text{hp}}. \tag{3.2}$$

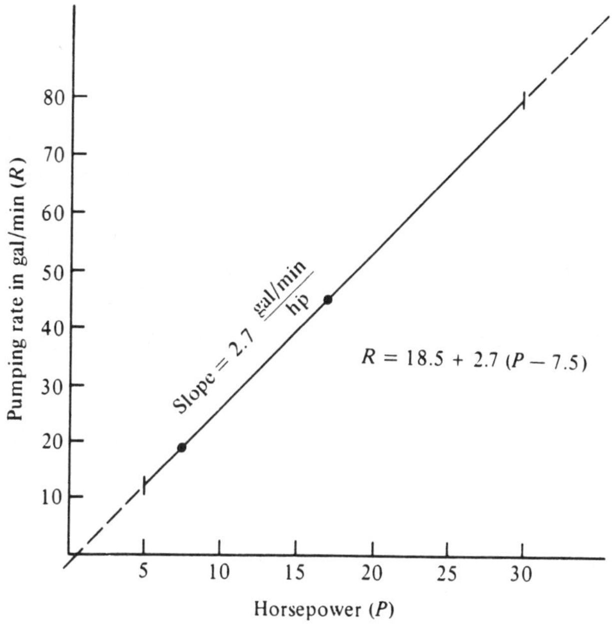

Figure 3.1 Diesel water pumps (5–30 hp)

So

$$\text{change in } R = 2.6842 * \text{change in } P. \tag{3.3}$$

We shall compare the 10-hp pump we are interested in with the 7.5-hp pump we know about. The change in P is $10 - 7.5$ hp, while the change in R is $R(10) - 18.5$ gal/min. Thus,

$$R(10) - 18.5 = 2.6842 * (10 - 7.5),$$

so

$$\begin{aligned} R(10) &= 18.5 + 2.6842(10 - 7.5) \\ &= 25.210\dot{5}\ \frac{\text{gal}}{\text{min}} \\ &\sim 25.2 \text{ gal/min}. \end{aligned} \tag{3.4}$$

Notice that in Equation 3.4 we used the value 2.6842 . . . which we found for the slope before rounding to two significant digits. That prevents rounding errors from accumulating. If we had used 2.7 instead, and then rounded again when we were done, we would have found a different (and wrong) answer:

$$R(10) = 18.5 + 2.7(10 - 7.5)$$
$$= 25.25$$
$$\sim 25.3 \text{ gal/min.}$$

The general rule is: *Use numbers* as accurate as possible in intermediate calculations.

The easiest way to save the value 2.6842 . . . in this example is to store it in the calculator's memory when you first find it. For the rest of this example we will show both the rounded, reported answers and some of the extra digits your calculator shows.

The second term on the right in the first line of Equation 3.4 is the additional pumping capacity caused by the extra 2.5 hp. The point on the graph in Figure 3.1 that corresponds to the 10-hp pump is between the points corresponding to the 7.5- and 18-hp pumps.

When the point we wish to study lies beyond the two given points the algebra is the same, but the process is called *extrapolation*. We extrapolate to study a 25-hp pump. When we compare it to the 17-hp pump and use Equation 3.3 we see that

$$R(25) - 44 = 2.6842 * (25 - 17).$$

So

$$R(25) = 44 + 2.6842(25 - 17) \tag{3.5}$$
$$= 65.4736 \ \frac{\text{gal}}{\text{min}}$$
$$\sim 65.5 \text{ gal/min.}$$

This answer is less reliable than the answer in Equation 3.4 to the interpolation problem–extrapolation is riskier. When you move beyond the data points the assumption that the dependence is linear is less reasonable than when you stay between.

Our final problem involves another interpolation. To see how powerful a pump is required to move 30 gal/min, we use Equation 3.3 in a slightly different way. Now P is unknown, and $R = 30$ gal/min. Hence, comparing with the 7.5-hp pump, we find that

$$30 - 18.5 = 2.6842(P - 7.5).$$

Thus

$$P - 7.5 = \frac{30 - 18.5}{2.6842},$$

and

$$\begin{aligned} P &= 7.5 + \frac{30 - 18.5}{2.6842} \\ &= 11.7843 \text{ hp} \\ &\sim 11.8 \text{ hp.} \end{aligned}$$

We could also have solved the problem by comparing this pump with the 17-hp pump, carefully watching minus signs. Equation 3.3 says

$$30 - 44 = 2.6842(P - 17),$$

so

$$P - 17 = \frac{30 - 44}{2.6842},$$

and

$$\begin{aligned} P &= 17 + \frac{30 - 44}{2.6842} \\ &= 11.7843 \text{ hp} \\ &\sim 11.8 \text{ hp.} \end{aligned}$$

This gives the same answer.

EXAMPLE 2. DIESEL WATER PUMPS, CONTINUED

Next we extend the method of interpolation to produce a formula expressing R in terms of P that still avoids mention of the intercept. We do this by comparing a P-hp pump with the 7.5-hp pump. The change in power is $P - 7.5$ hp; the corresponding change in pumping rate is $R - 18.5$ gal/min. Therefore, substituting in Equation 3.3, we get

$$R - 18.5 = 2.6842(P - 7.5),$$

and

$$R = 18.5 + 2.6842(P - 7.5) \\ \sim 18.5 + 2.7(P - 7.5). \tag{3.6}$$

The second term on the right in Equation 3.6 is the additional pumping power produced by the extra $P - 7.5$ hp.

If we were to multiply out the parentheses in Equation 3.6 and collect terms we would find that

$$\begin{aligned} R &= 18.5 + 2.6842P - 2.6842 * 7.5 \\ &= (18.5 - 2.6842 * 7.5) + 2.6842P \\ &= -1.6316 + 2.6842P \\ &\sim -1.6 + 2.7P \text{ gal/min.} \end{aligned} \tag{3.7}$$

The -1.6 gal/min is the intercept, which earlier we estimated at -2 gal/min, but which we considered to be meaningless. Equations 3.6 and 3.7 are algebraically equivalent, but the former makes more sense than the latter in the context of the problem. Since we know the dependence is linear only in the range from 5 to 30 hp, we should avoid mentioning values outside that range if possible.

EXAMPLE 3. UNITED STATES POPULATION

Table 1.2 on page 19 shows the United States population as counted by the census every 10 years from 1800 to 1960. Estimate the population in 1932, when Franklin Roosevelt was first elected president.

We must read between the lines of the table. The graph in Figure 3.2 shows the five points corresponding to the years 1910–1950. (Notice the breaks in the vertical and horizontal scales.) The points do not lie on a straight line because the population did not grow linearly with time. Nevertheless, to estimate the 1932 population we will suppose the growth was linear for the ten-year period 1930–1940. Letting P stand for the population (in millions) and Y for the year, we compute the slope for that period:

$$\begin{aligned} \frac{\text{change in } P}{\text{change in } Y} &= \frac{131.7 - 122.8}{10} \\ &= 0.89 \text{ million people/year.} \end{aligned}$$

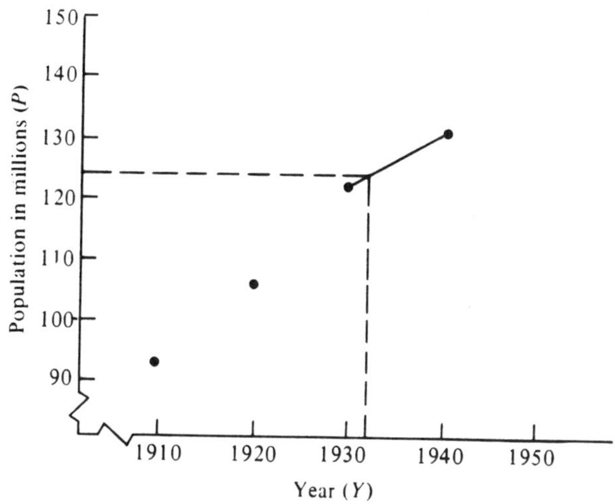

Figure 3.2 United States Population in 10-year intervals

Then,

$$\text{change in } P = 0.89(\text{change in } Y),$$

so for the two-year period from 1930 to 1932,

$$\text{change in } P = 0.89 * 2 = 1.78 \text{ million.}$$

Hence our interpolated estimate for the 1932 population is 1.78 more than the 1930 population, or 124.58 million, which we round to 124.6 million.

We can extrapolate backward, too, from the population in the years 1800 and 1810. The slope for that period is

$$\frac{7.2 - 5.3}{10} = 0.19 \, \frac{\text{million people}}{\text{year}}.$$

Thus the population in 1770, $P(1770)$, can be estimated to have been

$$P(1770) = P(1800) + (-30 \text{ years}) * 0.19 \, \frac{\text{million people}}{\text{year}}$$
$$= -0.4 \text{ million people.}$$

This impossible result shows again that extrapolation is not always safe, and must be justified each time. Pursue algebraic solutions only as long as your answers make sense.

PROBLEMS

1. Extrapolate in Example 1 (Diesel water pumps) to find the rate at which a 5-hp pump moves water. Be careful with minus signs.

2. In Example 1 recalculate the pumping rate for a 25-hp motor, using the rounded value of 2.7 gal/min instead of the correct value 2.6842 . . . gal/min. Compare your answer with the answer in the text in Equation 3.5.

*3. A 25-pound bag of dog food costs $4.79; a 10-pound bag costs $2.19. Interpolate and extrapolate to guess at the cost of a 15- and a 30-pound bag.

4. In 1906 the ore in the Bingham Canyon mine was 1.93 percent copper. In 1980 it was 0.60 percent copper. It will no longer be economical to mine the ore when its copper content drops to 0.40 percent. Extrapolate to estimate when that will occur.

*5. Life expectancy for United States males was 65 in 1950 and 68 in 1970. What will life expectancy be in the year 2000? When will it reach 101 years? When 200 years? Comment on the reasonableness of your answers.

6. In America men's shoe sizes are in the range 1–16. The corresponding range in Europe is 30–50. Write formulas analogous to Equation 3.6 for converting American shoe sizes to European and vice versa.

7. A Goodblimp outlet finds that when they set the price of a tire at $35 they sell 250 per month, while at a selling price of $45 they sell only 130 per month.

   (a) How many could they sell if they charged $40 per tire?

   (b) Interpret the slope you calculated in order to interpolate.

   (c) What assumption have you made in order to answer the question in (a)?

   (d) Under the same assumption, how many tires could they give away?

*8. When large eggs cost 95 cents per dozen the A&P sells 455 thousand dozen per day. When the cost is $1.15 per dozen, the demand is reduced to 320 thousand dozen per day. Suppose that demand depends linearly on price.

(a) Interpolate to find the demand when the price of a dozen eggs is $1.06. Extrapolate to find the demand when the price is 92 cents.

(b) At what price does the demand for eggs drop to 0? How many dozen eggs could the A&P give away? Are these reasonable extrapolations?

9. Extrapolate from Table 1.2 on page 19 to estimate the United States population in 1970 and in 1980, using the 10-year period from 1950 to 1960 to find the rate of population growth. Then find the correct figures for those years, and compare them with your extrapolations. (See Problem 20 of Chapter 1.)

10. Table 3.1 gives data from a milk carton. It shows how the length of time until the milk sours depends on the temperature at which it is stored.

(a) Draw a graph exhibiting the data.

(b) Interpolate to estimate the life of the milk at 36°F and at 53°F. Extrapolate to estimate its life at 80°F; at 90°F; and at 0°F.

(c) Discuss the usefulness of each of the estimates obtained in (b).

Table 3.1 Souring time for stored milk

Temperature (°F)	Storing time (days)
70	1/2
60	1
50	2
45	5
40	10
32	24

*11. The annual premium for $60K worth of homeowner's insurance is $437. The premiums for $80K and $100K are $608 and $792, respectively. (Remember, K means *thousand*.)

(a) Does the premium depend linearly on the value of the insurance?

(b) What is a reasonable estimate for the premium on a $95K homeowner's insurance policy? On a $115K policy?

12. Here is more of the quotation that opens this chapter:

> In the space of one hundred and seventy-six years the Lower Mississippi has shortened itself two hundred and forty-two miles. That is an average of a trifle over one mile and a third per year. Therefore, any calm person, who is

not blind or idiotic, can see that in the old Oölitic Silurian Period, just a million years ago next November, the Lower Mississippi River was upward of one million three hundred thousand miles long, and stuck out over the Gulf of Mexico like a fishing-rod. And by the same token any person can see that seven hundred and forty-two years from now the Lower Mississippi will be only a mile and three-quarters long, and Cairo and New Orleans will have joined their streets together, and be plodding comfortably along under a single mayor and a mutual board of aldermen. There is something fascinating about science. One gets such wholesale returns of conjecture out of such a trifling investment of fact.

Check Mark Twain's arithmetic. Then figure out how long the Lower Mississippi was when he wrote about it.

*13. Round each of the following numbers to 3, 2, and 1 significant digits.

(a) 3.2341 (b) 323 410 (c) 0.0032341
(d) 8.7678 (e) 876 780 (f) 0.0087678
(g) 53.012 (h) 53.092 (i) 53.992

14. (a) Round 746 to one significant digit.
(b) Round 746 to 2 significant digits, and then round the answer to one significant digit.
(c) Compare your answers to (a) and (b). What error is this problem designed to help you avoid?

15. (a) Calculate:

$$\frac{17.412 - 12.881}{0.62983 - 0.21309}.$$

Round the answer to 3 significant digits.
(b) Round the data in (a) to 3 significant digits and then perform the calculation. Round the answer to 3 significant digits and compare it with the answer to (a).

16. (a) Find the value of 9/87 654 with your calculator. Do not turn the calculator off. Write down the answer exactly as it is shown in the display. Now multiply the answer by 87 654. Do you get 9.000000 . . .?
(b) Next, clear the calculator and enter the number you wrote above as 9/87 654. Multiply that by 87 654. The answer is probably not exactly 9. That is because the answer the calculator showed was rounded to fit within the display. A good machine can remember more decimal places than it is able to

show, and used those places when you multiplied by 87 654 the first time. The second time, when you reentered the display you'd copied, the calculator had no access to the extra digits. The moral is: Never round in the middle of a calculation. Don't copy numbers from the display only to reenter them later on.

(c) If you would like to see if your calculator stores digits it doesn't show, calculate 9/87 654 again. Now multiply the result by 10 000.

CHAPTER 4

Proportion

Multiplication is vexation,
Division is as bad;
The rule of three doth puzzle me,
And practice drives me mad.
– Anonymous (Elizabethan manuscript)

Recognizing direct proportion in various disguises
Proportionality constants / Rates and slopes with their units
Percentage / Unit conversion and the metric system
Units for areas and volumes / Joint and inverse variation

EXAMPLE 1. SPEED

If you drive at 55 miles per hour, how is the distance you travel related to the time you spend? How far do you go in 2 hours and 15 minutes? How long will it take you to drive 295 miles?

The variables in this problem are the distance D, measured in miles, which depends on the time T, measured in hours. The equation relating them is

$$D = 55T. \tag{4.1}$$

Two hours and 15 minutes is 2.25 hours (not 2.15 hours). Substituting in Equation 4.1 gives

$$D = 55 * 2.25 = 123.75 \sim 124 \text{ miles.}$$

Figure 4.1 shows the graph of the dependence of D on T. It is linear, with a slope of 55 miles per hour. The intercept is special: it is

0 miles. That is visible in both Figure 4.1 and Equation 4.1. The intercept is 0 because when you travel for 0 hours you've gone 0 miles. There is no overhead, no membership fee, no set-up time, as in the examples and problems in the preceding chapters.

Linear dependence when the intercept is 0 is so important that special words are used to describe it. In this example we say that D *is proportional to* or *directly proportional to* T, or that D *varies directly with* T. The slope of 55 mi/hr is called the *proportionality constant.*

In general, a quantity Q is proportional to another, P, when there is some constant k such that

$$Q = kP. \tag{4.2}$$

The units of the proportionality constant k are

$$\text{units of } k = \frac{\text{units of } Q}{\text{units of } P}.$$

Observe that when Q is proportional to P we can solve Equation 4.2 for P in terms of Q and conclude that

$$P = \frac{1}{k}Q,$$

so that P is proportional to Q with the proportionality constant $1/k$. In

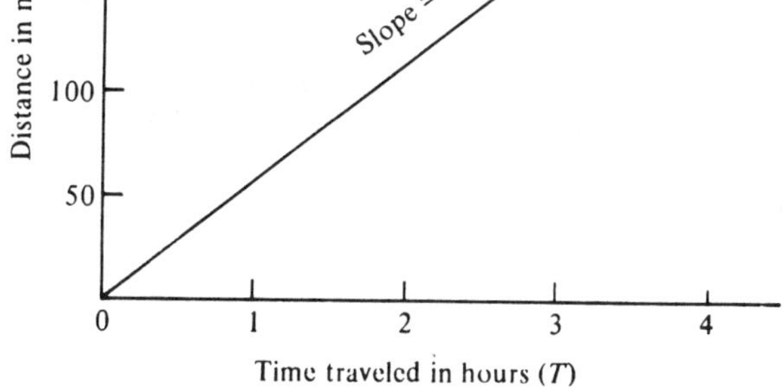

Figure 4.1 Travel at a steady speed

the example we are working, T is proportional to D. The proportionality constant is

$$\frac{1}{\frac{55 \text{ miles}}{\text{hour}}} = 0.0182 \frac{\text{hours}}{\text{mile}}.$$

When we divide each side of Equation 4.1 by 55 miles/hour to produce

$$T = 0.0182D, \tag{4.3}$$

we invert the units fraction miles/hour to hours/mile, and we invert the number 55 (with the [1/*x*] key) to 0.0182. Substituting $D = 295$ in Equation 4.3 tells us that $T = 5.36$ hours, or about 5 hours 22 minutes.

Here are two more examples of direct proportionality:

(a) When gasoline costs $1.64 per gallon then what you pay, P (measured in dollars), is proportional to the amount you buy, A (measured in gallons). The slope or proportionality constant is 1.64 $/gal:

$$P = 1.64A.$$

Equivalently, A is proportional to P; here the constant of proportionality is 0.610 gal/$.

(b) Your week's income, W (measured in dollars), is directly proportional to the number of hours you work:

$$W = rH.$$

Here the proportionality constant r is your hourly rate of pay. The units of r are $/hour.

EXAMPLE 2. THE RULE OF THREE

Suppose 143.4 gallons of home heating oil cost $170.86. How much will 200 gallons cost? How much oil will $200 buy?

You may have learned to answer questions like these in elementary school. We attack them here in a new, systematic way so that we can work more difficult ones later. First, we recognize the problem as one in direct proportion. The cost C, in dollars, is proportional to the number of gallons G purchased. The units of the proportionality constant will be $/gal. Since that constant is not given we will have to find it. Give it a name: k is traditional. Then

$$C(\$) = k\left(\frac{\$}{\text{gal}}\right) * G(\text{gal}). \tag{4.4}$$

We know that $C = \$170.86$ when $G = 143.4$ gal; we can substitute those values in Equation 4.4 to get

$$170.86\$ = k\,\frac{\$}{\text{gal}} * 143.4 \text{ gal}. \tag{4.5}$$

Hence

$$k = \frac{170.86\$}{143.4 \text{ gal}} = 1.1915\,\frac{\$}{\text{gal}} \sim 1.19 \text{ \$/gal}. \tag{4.6}$$

Now we can answer the questions asked. To find C when $G = 200$ gal, use Equation 4.4 with our newfound value for k:

$$C = 1.1915\,\frac{\$}{\text{gal}} * 200 \text{ gal} = \$238.30.$$

This is the cost of 200 gallons rounded to the nearest cent. Notice that we did not round the value of k to 1.19 \$/gal while we were doing the problem. If we had used 1.19 \$/gal instead of the true value we stored in the calculator we would have found only a rough cost of \$238.00 for 200 gallons of oil.

How much oil will \$200 buy? We must find G when $C = \$200$.

$$200\$ = 1.1915 \text{ \$/gal} * G \text{ gal},$$

and thus

$$G = \frac{200\$}{1.1915 \text{ \$/gal}} = 167.9 \text{ gal}. \tag{4.7}$$

Notice how the units cancel in Equation 4.7: $\$/(\$/\text{gal}) = \text{gal}$.

The point of this example is to display the method embodied in Equations 4.4, 4.5, and 4.6 for finding the proportionality constant. In Equation 4.6 we essentially found a slope. It looks simpler than the earlier slope computations in Chapters 2 and 3 because, in direct-proportion problems like this one, the intercept is known to be 0. Thus when the data, 143.4 gallons cost \$170.86, are compared with the intercept data, 0 gallons cost \$0, the *changes* in C and G are easy to compute:

$$143.4 - 0 = 143.4,$$

and

$$170.86 - 0 = 170.86,$$

so Equation 4.6 is in fact finding

$$\frac{\text{change in } C}{\text{change in } G}.$$

When you first learned to work these problems, slopes were not mentioned. You probably reasoned as follows: If x stands for the cost of 200 gallons, then

$$x \text{ is to } 200 \text{ as } 170.86 \text{ is to } 143.4, \tag{4.8}$$

or

$$\frac{x}{200} = \frac{170.86}{143.4}. \tag{4.9}$$

Cross-multiplying and solving for x gives

$$x = 200 * \frac{170.86}{143.4} = \$238.30.$$

That quick, correct solution is a variant of an old method called *the rule of three*, because three of the four numbers in the proportion 4.8 are known. In the rule of three the proportionality constant k enters unnoticed on the right side of Equation 4.9. We have introduced a new method in order to make that proportionality constant visible.

EXAMPLE 3. PERCENTAGE

The sales tax in Massachusetts is 5 percent. Write that number without using the percent symbol (%), and study its units.

When you buy something, the sales tax T you pay depends on the dollar value P of your purchase according to the rule

$$T = 0.05P. \tag{4.10}$$

This says in words "T is 5 percent of P," because *percent* means *divide* *by 100*, and the word *of* almost always means *multiply*. Thus 5% means 5/100 or 0.05, so "5% of P" means "0.05 * P."

Since Equation 4.10 is linear with an intercept of 0, it represents direct proportion. What are the units of the proportionality constant 0.05? To find out, rewrite the equation with units added:

$$T(\$\text{ of tax}) = 0.05 * P(\$\text{ of purchase}). \tag{4.11}$$

Therefore

$$0.05 = \frac{T(\$\text{ of tax})}{P(\$\text{ of purchase})}, \tag{4.12}$$

and the units of the proportionality constant are

$$\frac{\$\text{ of tax}}{\$\text{ of purchase}}. \tag{4.13}$$

If we choose to regard ($ of tax) and ($ of purchase) as the same kind of dollars, then we can cancel them in Equation 4.12, and the proportionality constant becomes 0.05, a pure number with no units.

Working with percentages is easier if you pay attention to two potential sources of confusion. The first is the apparent lack of units. In this example we have discovered that the number 0.05 really measures dollars (of tax) per dollar (of purchase). The second is that most percentages are numbers less than 1. I suspect that percentages were invented because people preferred the ease of working with whole numbers like 5, regarded as *out of 100*, to struggling with the fraction 5/100 or the decimal 0.05.

> Your calculator may have a [%] key. All it does is move the decimal point two places to the left. Experiment with it if you wish; we shall make no further reference to it in this book.

We shall practice more with percentages. For example, most wine is about 11.5% alcohol. To translate this into a sentence involving direct proportion, start by labeling the variables. Let W be the quantity of wine, measured in any convenient units–say ounces. Then let A be the quantity of alcohol (also measured in ounces) in the wine. The proportionality constant is 11.5%, or the fraction 11.5/100, or, converted to a decimal: 0.115 ounces of alcohol per ounce of wine. Thus

$$A = 0.115W. \tag{4.14}$$

The Bingham Canyon mine yields ore that is about one half of one percent copper. (See Problem 4 of Chapter 3.) Since

$$\frac{1}{2}\% = 0.5\% = 0.005,$$

it follows that

$$C = 0.005R. \tag{4.15}$$

Here C is the number of pounds of copper contained in R pounds of ore. The units of the proportionality constant are pounds of copper/pound of ore. But notice that Equation 4.15 also gives the number of *tons* of copper in R *tons* of ore. That is why constants with no units are useful: as long as we use the same units for C and R, the numerical value of the proportionality constant is the same. The *proportion* of the ore that is copper is 0.005, or 0.5%.

EXAMPLE 4. UNIT CONVERSIONS AND THE METRIC SYSTEM

How many seconds are there in 6.5 minutes?

That's easy. Simply multiply 6.5 by 60 to find that there are 390. But in order to carry out more sophisticated conversions we shall examine this one more carefully. It is direct proportion in disguise. The 60 is really a proportionality constant of 60 seconds per minute:

$$S = 60M, \tag{4.16}$$

where S is the number of seconds and M the number of minutes.

To find the number of seconds in a year we construct the following chain of proportions:

$$60\,\frac{\text{seconds}}{\cancel{\text{minute}}} * 60\,\frac{\cancel{\text{minutes}}}{\cancel{\text{hour}}} * 24\,\frac{\cancel{\text{hours}}}{\cancel{\text{day}}} * 365.25\,\frac{\cancel{\text{days}}}{\cancel{\text{year}}} * 1\,\cancel{\text{year}}$$
$$= 31\,557\,600 \text{ seconds.}$$

The units minutes, hours, days, and year all cancel, leaving seconds alone on the right. In complicated problems, when it is hard to know whether to multiply or divide by the conversion factor, the units will tell you which to do.

Since unit conversions are problems in direct proportion they are easy to invert. For example, Equation 4.16 can be rewritten as

$$M = \frac{1}{60}S = 0.0167S.$$

This tells how to convert seconds into minutes. The units of the constant 0.0167 are min/sec.

Table 4.1 lists the conversion factors (proportionality constants) for converting between the metric and English systems. People who use the metric system regularly do not need these conversion factors. They have a direct sense of the sizes of the metric units. Our children may come to think of a centimeter as, say, the radius of a penny rather than as just under half an inch, but we must be able to convert from centimeters to inches and back.

Once you are used to the metric system all conversions within it are easy. You need not remember such conversion factors as 36 inches/yard, 5280 feet/mile, 32 ounces/quart and 16 ounces/pound; you just learn a few prefixes. Instead of multiplying or dividing by 36, 5280, 32, or 16, you just move a decimal point to the right or to the left. For example, the prefix *kilo* (symbol *k*), means 1000. Thus 1 kilogram equals 1000 grams. To convert from kilograms to grams, multiply by the proportionality constant 1000 g/kg. For example, you would convert 2.3 kilograms to grams like this:

$$2.3 \text{ kg} = 2.3 \cancel{\text{kg}} * 1000 \frac{\text{g}}{\cancel{\text{kg}}} = 2300 \text{ g}.$$

Table 4.1 Introduction to the metric system

Metric unit	Abbreviation	Casual conversion	Factor
Length			
meter	m	width of a doorway, 10% more than a yard	0.9144 m/yd
centimeter	cm	radius of a penny, just under half an inch	2.540 cm/in
kilometer	km	10-minute brisk walk, just over a half mile	1.609 km/mi
Volume			
liter	L	a little more than a quart	0.9463 L/qt
milliliter	mL	eyedropperful	29.57 mL/oz
Weight			
gram	g	half a sugar cube	28.35 g/oz
kilogram	kg	just over two pounds	0.4536 kg/lb

Table 4.2 Some metric prefixes

Prefix	Abbreviation	Meaning
kilo	k	1000 *
centi	c	0.01 *
milli	m	0.001 *

In other words you move the decimal point 3 places to the right. The identical argument would be used to convert 2.3 kilometers to 2300 meters.

To convert from grams to kilograms, use the reciprocal constant, 0.001 g/kg. In other words you move the decimal point 3 places to the left. Table 4.2 gives a few of the common metric prefixes. From it we see that 1 cm = 0.01 m, and 100 cm = 1 m.

Let us combine the techniques we've just learned, and convert a speed of 27 miles per hour to its equivalent in meters per second. If we multiply 27 mi/hr by the correct conversion factor for km/mi then the units *miles* will cancel, leaving km/hr. Then we multiply by the conversion factor for m/km to cancel the *km* and leave m/hr. Finally, we multiply by the factor for hr/min and min/sec so that we are left with an answer in m/sec. All this is easier to do than to explain:

$$
\begin{aligned}
27\,\frac{\text{mi}}{\text{hr}} &= 27\,\frac{\cancel{\text{mi}}}{\cancel{\text{hr}}} * 1.609\,\frac{\cancel{\text{km}}}{\cancel{\text{mi}}} * \frac{1000\text{ m}}{\cancel{\text{km}}} * \frac{1\ \cancel{\text{hr}}}{60\ \cancel{\text{min}}} * \frac{1\ \cancel{\text{min}}}{60\text{ sec}} \\
&= \frac{27 * 1.609 * 1000}{60 * 60}\ \frac{\text{m}}{\text{sec}} \\
&= 12.0675\,\frac{\text{m}}{\text{sec}} \\
&\sim 12\text{ m/sec.}
\end{aligned}
$$

EXAMPLE 5. AREAS

What is the area of a wall that is 2.5 m high and 4.3 m long? Express the answer both in square meters and in square yards.

That is the sort of question which can come up when you must decide how much paint or wallpaper to buy. To answer it, remember that the area of a rectangle is its height times its length, or

$$2.5\text{ m} * 4.6\text{ m} = 11.5\text{ m} * \text{m} = 11.5\text{ m}^2. \tag{4.17}$$

Pay particular attention to the units. They are meters * meters, or *square meters* (m^2).

To convert from m * m to yd * yd we must multiply *twice* by the conversion factor for yd/m. Table 4.2 tells us there are 0.9144 m/yd; it is the reciprocal of that factor we need.

$$
\begin{aligned}
11.5 \text{ m}^2 &= 11.5 \text{ m}^2 * \left(\frac{1 \text{ yd}}{0.9144 \text{ m}}\right)^2 \\
&= \frac{11.5}{0.9144^2} \text{ yd}^2 \qquad (4.18) \\
&= 13.8 \text{ yd}^2.
\end{aligned}
$$

EXAMPLE 6. JOINT VARIATION

How much interest do you pay on a loan?

Suppose you pay 17.6% annual interest on a loan for a new car. Then in a year the interest I (in \$) is directly proportional to the amount borrowed—the principal P (in \$):

$$I = 0.176P. \qquad (4.19)$$

The units of the proportionality constant 0.176 are dollars of interest per dollar of principal. In general, when the interest rate is r (usually given as a percentage but always written as a decimal) we have

$$I = rP. \qquad (4.20)$$

Before we can figure out how much interest you pay we must know how long you hold the bank's money. Equations 4.19 and 4.20 are based on a one-year loan. But the longer you keep the money, the more interest you must pay. The amount of interest is directly proportional to the duration T (measured in years) of the loan:

$$I = 0.176PT, \qquad (4.21)$$

or, in general,

$$I = rPT, \qquad (4.22)$$

when the interest rate is r.

We describe Equation 4.22 by saying that I is *jointly proportional* to r, P, and T. If any two of the three are fixed, I is proportional to the third. Thus if the time period T and the interest rate r are known, the interest I is proportional to the principal P—the more you borrow, the more interest you pay. If, on the other hand, you borrow a fixed amount, say \$2500, at a fixed rate, say 11%/year, then your interest I is proportional to T, the length of time you keep the loan:

$$I = 0.11 * 2500 * T = 275T.$$

The units of the proportionality constant 275 are

$$\frac{\text{\$ of interest}}{\text{year}}.$$

If the time T and principal P are fixed, the interest I is proportional to the interest rate r.

In each case the units work out nicely if we remember that the units of the interest rate r are not just

$$\frac{\text{\$ of interest}}{\text{\$ of principal}},$$

but

$$\frac{\frac{\text{\$ of interest}}{\text{\$ of principal}}}{\text{year}}.$$

Then the units in Equation 4.22 are

$$\text{\$ of interest} = \frac{\frac{\text{\$ of interest}}{\text{\$ of principal}}}{\text{year}} * \text{\$ of principal} * \text{years}. \quad (4.23)$$

A word of warning to the borrower: You do not usually pay back a loan all at once with interest at the end of the time period. Instead, you pay some principal and some interest every month. Often the true interest rate is then larger than the advertised rate. Truth-in-lending laws generally require the lender to tell you the true rate of interest, if only in fine print.

A word of warning to the calculator: One of the fraction bars in the compound fraction in Equation 4.23 must be longer than the other in order to distinguish the order in which the calculations with the units are to be performed. For a numerical illustration, see Problem 41.

EXAMPLE 7. INVERSE PROPORTION

How long does it take to drive 320 miles?

We know from Example 1 that the distance you travel at a fixed speed is directly proportional to the time; the speed is the proportionality constant. This example differs. The distance, 320 miles, is fixed, while the speed V (in miles per hour) is one of the variables. (We use V rather than S for the variable here for two reasons: a handwritten S is easily confused with a 5; and V stands for *velocity*.) The time T (in hours) you need to drive the 320 miles depends on V. Since

$$320 = VT,$$

the dependence of V on T is given by

$$T = \frac{320}{V}. \tag{4.24}$$

For example, when $V = 55$ mi/hr, $T = 5.8$ hours. When $V = 60$ mi/hr, $T = 5.3$ hours. The faster you drive, the less time it takes. The dependence of T on V is graphed in Figure 4.2. It is *not* linear. When V is very small—say 2 mi/hr—T is very large: 640 hours. When V

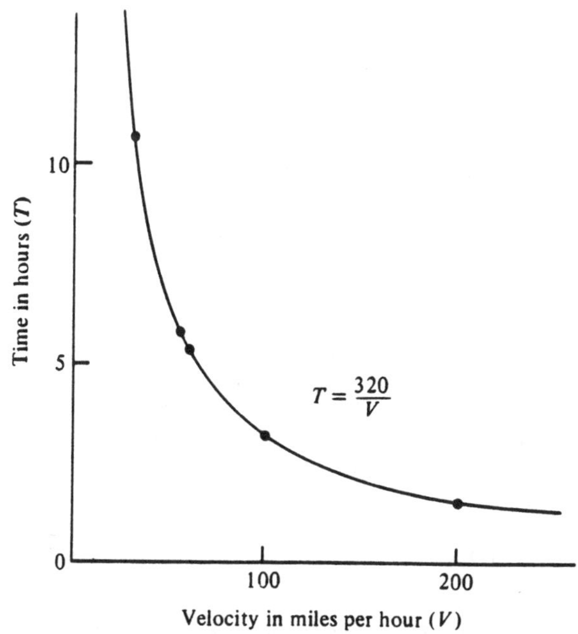

Figure 4.2 How time depends on rate

is outrageously large—say 800 mi/hr—then T is very small: 0.4 hours. We describe this relationship by saying that T is *inversely proportional to* or *varies inversely with* V: there is a constant k such that

$$T = \frac{k}{V}.$$

The units of k are

$$(\text{units of } T) * (\text{units of } V).$$

In this example, $k = 320$ miles.

PROBLEMS

Whenever it is appropriate, identify the variables and their units, the proportionality constant with its units, and round numerical answers to the correct number of significant digits.

*1. I cycle 4.8 miles in 23 minutes. How far can I go in an hour? How long does it take me to ride a mile? How long would it take me to ride 150 miles?

2. The director of the cafeteria at UMass/Boston knows it cost $28 700 to serve 31 250 meals in March. She expects to serve approximately 26 000 meals in April. How much should she anticipate spending?

3. Bradlee's advertises 15- and 18-inch gold chains on sale at 99 cents per inch.
   (a) Set up the implicit problem in direct proportion, and find the cost of each length of chain.
   (b) The regular price is $1.34 per inch. Show that the amount you *save* is proportional to the length of the chain.

4. Liquid penicillin contains 250 000 units (of penicillin) per milliliter of solution. How many milliliters should be prescribed for a patient who needs 6 million units?

*5. The proportion of albinos in the rabbit population is 0.013. What is the percentage of albinos? How large a population will contain 380 albinos?

6. When refined in the laboratory, a 167-g ore sample produced 2.16 g of iron. What percentage of the ore is iron? How much iron

will 1000 kg of ore yield when it is refined commercially? How much ore must be processed to produce 50 kg of iron?

7. (a) A six-pack of 6-ounce cans of tomato juice sells for $1.19; the 1-quart 14-ounce can sells for 95 cents. Compare these prices by finding the appropriate proportionality constants.
   (b) When you are next in a supermarket look at the *unit pricing* information displayed on the shelves. (Some states require that stores provide unit prices.) Copy a few. Explain why they are statements about direct proportion.

*8. Economic studies have shown that women's salaries are about 59 percent of men's for comparable work. A woman executive earns $32 500 per year. How much would she be earning for that work if she were male?

9. If you swallow twice each minute, how often do you swallow in a week?

10. How many meters are there in a mile? In a marathon? How many liters are in a gallon? How many cm in a km? Km in a cm?

*11. I ran a 10-km race in 49 minutes 11 seconds. Did I run 8-minute miles?

12. Convert 60 mi/hr to ft/sec; to km/hr; to m/sec.

13. When a stone is dropped from a bridge it falls faster and faster. For a while, its speed V (in ft/sec) depends on the time t (in seconds) it has been falling according to the formula

$$V = 32t.$$

   (a) How fast is it falling after 2.4 seconds?
   (b) How long until it is falling at 100 ft/sec?
   (c) What are the units of the proportionality constant 32?

*14. When the Pope visited Boston several years ago the police spent $80 000 for 40 miles of rope to control the crowds on the parade route. What problem in direct proportion is implicit here? What is the proportionality constant?

15. In October 1981 a French franc was worth $0.18, and coffee cost $2.95/pound. A friend from France would like you to convert that price to francs/kg. Please oblige. What is the current value of the franc in dollars? Why would this problem be more confusing if your friend were English? How did the English monetary unit get its name?

*16. A compact car can go 256 miles on 10.4 gallons of gas. Express the fuel economy in mi/gal and in gal/mi. How far can it go on 50 gallons of gas? How much gas is needed to travel 400 miles? Express the fuel economy in L/km and in km/L. Answer the same questions for your car. That will require collecting some data.

17. In baseball, the number of hits a batter makes is directly proportional to the number of times officially at bat. The proportionality constant is called the batting average. What are its units? If a player with a batting average of .286 has 320 official times at bat in a season how many hits would this represent?

18. Wine is 11.5% alcohol. How much alcohol is there in a fifth of wine? (A *fifth* is a fifth of a gallon.) This problem is easier in the metric system. How much alcohol is there in a liter of wine?

19. The auto excise tax varies directly with the value of the car. My old car was valued at $300; I paid $16.50 in tax. What is the tax rate? What are its units? How much tax will I have to pay on a new $8500 car?

20. Property tax is directly proportional to assessed valuation. A house assessed at $51 300 is taxed $1066.

(a) If property taxes are limited to two and one half percent of assessed valuation, is this rate legal?

(b) What are the taxes on a house assessed at $155 000?

21. The new Coke bottle is advertised as "2 liters–5% more than 64 ounces." What percentage more than 32 ounces is 1 liter?

*22. In 1978, 1 813 000 fans spent $3 126 000 for beer and hot dogs at Fenway Park, whose capacity is 31 000 people. How much money did the concessionaires fail to collect because (unfortunately) five post-season games were not played there?

*23. A 4-liter can of paint covers 22 square meters. How much paint will you need to paint the walls of a room that is 5 m long, 4 m wide, and 3.2 m high?

24. A map is drawn to the scale 1:4800. If a line on the map is 7 cm long, how long is the actual distance? (In km? In mi?) If a region has an area of 4 cm^2 on the map, what is the area of the region? (Answer in both metric and English units.)

25. A shipping carton measures 18″ long, 14″ wide, and 12″ deep. Find its volume in cubic inches, cubic feet, cubic centimeters, and cubic meters. How many small cubical boxes 10 cm on an edge will fit in the carton?

26. I have invented a new unit of time I call the *chron* (symbol c). There will be 10 chron per day.
    (a) Assuming that the daily chron count starts at midnight, at what time will my alarm (currently set for 6:30) go off?
    (b) How many kilochron are there in a year?
    (c) Find the factors for converting seconds to millichron and millichron to seconds.
    (d) Convert 30 mi/hr to the equivalent speed in km/c.
    (e) Discuss the advantages and disadvantages of this new unit.

27. Balsa wood, oak, and *lignum vitae* have densities of 8, 45, and 78 pounds per cubic foot, respectively.
    (a) Convert these densities to g/cm^3.
    (b) A substance floats when it is less dense than water. The density of water is 1 g/cm^3. (The gram was chosen as the metric unit of mass for just that reason.) Does all wood float?

*28. The length L of the empty part of a ballpoint pen cartridge is directly proportional to the length W of written line that has been produced. An experiment showed that when $L = 1.5$ cm, $W = 32$ m. Write the equation showing how W depends on L. Be sure to identify the units of the proportionality constant. How much writing can you do with a full 4-cm cartridge? Write an equation showing how the length F of the full part of the cartridge depends on W. Is F directly proportional to W?

*29. If you borrow $7000 from the bank for 9 months and pay $650 interest, what is the annual interest rate?

30. Suppose D dollars are invested in a savings account paying interest at a constant rate r. If *no interest is paid on the interest earned*, write the equation which shows that the total amount accumulated increases linearly with time.

31. Americans use domestic electric power at the rate of 1200 kWh (kilowatt-hours) per person per year. Write an equation showing how the domestic electric power use is jointly proportional to the number of users and the time under consideration. How much electricity is used in a city of 200 000 people in a month? How much in your city in a day?

32. Look back to Example 2 in this chapter, in which 143.4 gallons of home heating oil cost $170.86. What rate do you think the oil company was actually charging: 1.19 $/gal, 1.1915 $/gal, or 1.191492329 $/gal? Where did this last figure come from? Where did rounding take place?

*33. A sample of 325 silicon chips yielded 241 that passed inspection.

(a) How many good chips are there in a batch of 10 000?

(b) How many should be manufactured to get 650 good ones?

(c) Why are these answers probably only approximately correct?

34. Liz Ames's friend at the Metropolitan District Commission counted the fish in Jamaica Pond in this way: He caught 50 fish and tagged them. (The process doesn't harm the fish.) Then he threw them back, waited a while, and went fishing again. He caught 31 fish, and found tags on 4 of them. Approximately how many fish live in Jamaica Pond?

35. One draft of the 1980 census reported the black population of the United States as 26 488 218, 11.7% of the total. What is the total population? How many of the digits in the count of the black population do you think are significant?

36. A soccer player runs fairly steadily during a game for 40 minutes at an average speed of 5 miles per hour. If the field is 100 yards long and 50 yards wide how many times could the player run its length? How many laps? (A lap is once around the boundary of the field.)

37. A university advertises a faculty/student ratio of 1/14. Write the statement in direct proportion implied.

*38. A laboratory solution is 23% sulfuric acid.

(a) How much acid is there in 165 mL of the solution?

(b) How much water should be added to dilute the solution to one whose acid concentration is 14%? (In the laboratory be sure to add the acid to the water, not the water to the acid. Do you know why?) (This problem anticipates part of Chapter 5, Example 2.)

39. Suppose a quantity Q depends linearly on another quantity P.

(a) Show that Q need not be proportional to P.

(b) Explain why any change in Q is proportional to the corresponding change in P. (Compare with Problem 1 of Chapter 1.)

40. A kilowatt-hour (symbol kWh) of electrical energy will keep a 1000-W appliance running for 1 hour. The length of time you can run an appliance on 1 kWh of energy is inversely proportional to the rating of the appliance, in watts.

(a) How long can you run a 2250-W heater on 1 kWh of electricity? How long will 1 kWh keep a ½-watt nightlight burning?

(b) How many kWh are used keeping a 100-W bulb lit for a week? (This problem anticipates Example 5 of Chapter 8.)

41. Calculate (20/10)/2 and 20/(10/2). Which answer does your calculator give when you enter 20 [÷] 10 [÷] 2? The moral is: Avoid ambiguous expressions like 20/10/2!

42. The trucker's lament. How much extra time is spent driving from Salt Lake City to San Francisco when the driver observes the 55 mi/hr speed limit? Before you can answer the question you will have to supply two missing pieces of information.

***43.** The rate at which water flows through a horizontal pipe of moderate length is inversely proportional to the length of the pipe. Suppose a 10-foot pipe delivers a flow of 2 gal/min.

(a) Name the variables, with their units, and find the proportionality constant. What are its units? Write the resulting formula showing how flow rate depends on pipe length.

(b) What is the flow rate through a 20-foot pipe? Through a 3-foot pipe?

(c) How long a pipe produces a flow of 4 gal/min?

(d) Draw a graph of the relationship in (a).

44. The gas tank in most automobiles is designed to hold enough fuel for a trip of about 350 miles. Find two formulas showing how the size of the tank depends on the fuel economy of the car, one when the fuel economy is measured in mi/gal, and the other when measured in gal/mi. Draw the two graphs.

***45.** For a fixed quantity of lifting gas in a balloon, the volume is directly proportional to the absolute temperature and inversely proportional to the pressure.

(a) If the pressure is doubled from 1 atm to 2 atm while the temperature remains the same, what happens to the volume?

(b) If the absolute temperature is doubled while the pressure remains the same, what happens to the volume?

(c) If both the pressure and the absolute temperature are halved, what happens to the volume?

46. If a hen and a half lays an egg and a half in a day and a half how long will it take one hen to lay one egg? The answer to that old riddle is not *one day*. Explain why it is not. Calculate the production rate in terms of (eggs/hen)/day. Use that rate to answer the riddle correctly.

47. When I was in the second grade in PS 139 in Brooklyn the principal, Dr. Nathan Dickler, confounded the class by asking how much a horse weighed standing on two legs, if it weighed two thousand pounds when standing on four legs. That is not a problem in the *rule of three*. But my father and I each learned about that rule from Dr. Dickler. What *is* the answer to Dr. Dickler's question?

CHAPTER 5
Mixtures

Thou shalt not sow thy field with two kinds of seed; neither shall there come upon thee a garment of two kinds of stuff mingled together.

– *Leviticus* XIX, 19

Mixing gasohol, solutions, and investments / Problems in which the variables are proportions / Weighted averages

EXAMPLE 1. GASOHOL

Gasohol is a mixture of gasoline and alcohol that can be used instead of pure gasoline to run automobiles. Some day gasoline may cost 2.77 \$/gallon. Suppose that then alcohol costs 1.54 \$/gallon. How much will gasohol cost?

The question is clearly not yet complete. To answer it we still need to know how the mixture is made. If the gasohol is mostly gasoline, the cost will be nearly 2.77 \$/gallon. If it is mostly alcohol the cost will be just a little more than 1.54 \$/gallon. Now we can see what the variables are. The cost C of gasohol, measured in \$/gallon, depends on the way the mixture is prepared. To measure this let P represent the *proportion* of gasoline in the gasohol. For example, if the gasohol is 12% alcohol and 88% gasoline then $P = 0.88$. The units of P are (gal gasoline)/(gal gasohol). If you know the percentage of gasoline you find P by moving the decimal point two places to the left. A mixture that is half gasoline is 50% gasoline, so $P = 0.50$. A mixture for which $P = 0.651$ is 65.1% gasoline and 34.9% alcohol. The percentage of gasoline can be neither smaller than 0 nor larger than 100. Hence the proportion P of gasoline varies between 0 and 1. The value of P determines the gasohol's cost C.

Now that we know what the variables are, we can study how they are related. When P, the proportion of gasoline, is 0, the gasohol mixture is pure alcohol, and $C = 1.54$ \$/gallon. When $P = 1$ the gasohol is all gasoline, so $C = 2.77$ \$/gallon. What is C when $P = 0.88$, when the gasohol is 88% gasoline? Since the value of C is the cost of one gallon of the gasohol, we must find out how much gasoline and how much alcohol that one gallon contains. P tells us. Since the proportion of gasoline is $P = 0.88$, each gallon of gasohol contains 0.88 gallon of gasoline. That gasoline costs

$$2.77 \frac{\$}{\text{gal}} * 0.88 \text{ gal.}$$

Now how much alcohol is used, and what does it cost? Since one gallon of gasohol contains 0.88 gallon of gasoline, the other 1 − 0.88 gallon must be alcohol, which costs

$$1.54 \frac{\$}{\text{gal}} * (1 - 0.88) \text{ gal.}$$

The cost of the gasohol is the sum of the costs of its ingredients, and so when $P = 0.88$:

$$\begin{aligned} C &= 0.88 * 2.77 + (1 - 0.88) * 1.54 \\ &= 2.62 \frac{\$}{\text{gal}}. \end{aligned} \tag{5.1}$$

Notice that the cost is between the 1.54 \$/gallon for pure alcohol and the 2.77 \$/gallon for pure gasoline, but is much nearer the latter, because the gasohol is mostly gasoline.

Next we find a formula for C in terms of P, by following the same steps we took above for the particular value $P = 0.88$. One gallon of gasohol contains P gallons of gasoline; the other $(1 - P)$ gallon is alcohol. The P gallon of gasoline costs \$2.77 P; the $(1 - P)$ gallon of alcohol costs \$1.54$(1 - P)$. Since C is the total cost of one gallon of gasohol,

$$C = 2.77P + 1.54(1 - P). \tag{5.2}$$

Equation 5.1 results when we substitute $P = 0.88$ in Equation 5.2.

Although we discovered Equation 5.2 by considering one gallon of gasohol, its units are correct as they stand. C is measured in \$/(gal gasohol), P in (gal gasoline)/(gal gasohol), and $(1 - P)$ in (gal alcohol)/(gal gasohol). The units in Equation 5.2 are thus

$$\frac{\$}{\text{gal gasohol}} = \left(\frac{\$}{\text{gal gasoline}}\right)\left(\frac{\text{gal gasoline}}{\text{gal gasohol}}\right) + \left(\frac{\$}{\text{gal alcohol}}\right)\left(\frac{\text{gal alcohol}}{\text{gal gasohol}}\right).$$

Figure 5.1 shows graphically how C depends on P. We can exploit this graph and Equation 5.2 to answer several kinds of questions about the pricing of gasohol. If we know P, the proportion of gasoline, we substitute directly. Suppose, instead, we wish to know what the proportion of gasoline is in gasohol that sells for 1.80 \$/gallon. We start by solving Equation 5.2 for P in terms of C:

$$C = 2.77P + 1.54 - 1.54P = (2.77 - 1.54)P + 1.54,$$

so

$$(2.77 - 1.54)P = C - 1.54, \tag{5.3}$$

and

$$\begin{aligned} P &= \frac{C - 1.54}{2.77 - 1.54} = \frac{C - 1.54}{1.23} \\ &= \left(\frac{1}{1.23}\right)(C - 1.54) = 0.813(C - 1.54) \\ &= 0.813C - 1.25. \end{aligned} \tag{5.4}$$

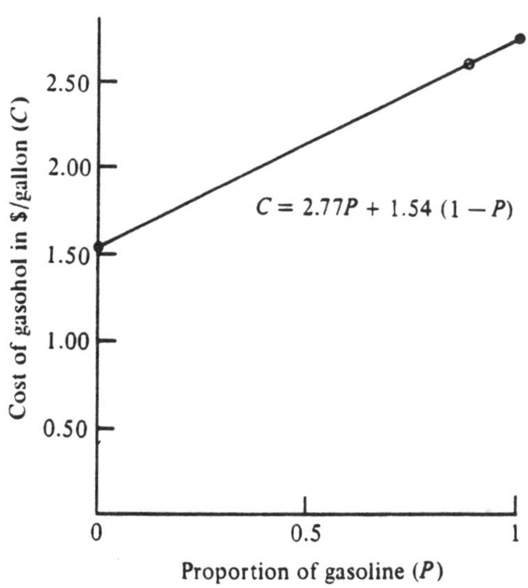

Figure 5.1 Finding the cost of gasohol

Then we substitute $C = 1.80$ in Equation 5.4 to conclude that $P = 0.211$. Hence gasohol that costs 1.80 \$/gallon is 21.1% gasoline and 78.9% alcohol.

In this problem it would have been easier to substitute $C = 1.80$ directly in Equation 5.3. But if we had many values of C for which we wanted to find P, it would make sense to change Equation 5.2 into Equation 5.4 first.

EXAMPLE 2. MIXED NUTS

No algebra book would be complete without a problem on the price of a mixture of peanuts and cashews. Peanuts cost 1.37 \$/lb. Cashews cost 3.13 \$/lb. How many pounds of each are required to prepare 70 pounds of a mixture that will sell for 1.95 \$/lb? If we have 200 lb of peanuts on hand, how many pounds of cashews will we need to make a mixture that sells for 2.75 \$/lb?

These questions each ask about *amounts* of peanuts or cashews. But before we can deal with amounts we must study the proportions. As in the gasohol example, the cost C (in \$/lb) of a mixture depends on the *proportion* P of, say, peanuts. (C also depends on the proportion of cashews. We could, if we wished, use that proportion instead as the independent variable.) P varies from 0 to 1. When $P = 0$, the mixture isn't really a mixture at all. It's all cashews and costs 3.13 \$/lb. When $P = 1$, the mixture is all peanuts and $C = 1.37$ \$/lb. The dependence of C on P is given by

$$C = 1.37P + 3.13(1 - P). \tag{5.5}$$

Notice that the coefficient of P, the proportion of peanuts, is the price of the peanuts. Solving Equation 5.5 for P in terms of C gives

$$P = 1.79 - 0.57C. \tag{5.6}$$

Now we are ready to answer the particular questions asked. If a mixture is to sell for $C = 1.95$ \$/lb, we find the proportion of peanuts by substituting in Equation 5.6:

$$P = 1.79 - (0.57 * 1.95) = 0.67.$$

Thus the proportion of peanuts is 0.68, or 68%. The mixture will have to be approximately 2/3 peanuts. This is indicated in Figure 5.2, in which 1.95 is about 1/3 of the way from 1.37 to 3.13. Since 70 pounds of mix is required, we will need

$$0.67 \frac{\text{lb peanuts}}{\text{lb mix}} * 70 \text{ lb mix} = 47 \text{ lb peanuts.}$$

The other 23 pounds should be cashews.

The next question, which asks how many pounds of cashews we should add to 200 lb of peanuts, is trickier. The price for the mixture is to be $C = 2.75$ \$/lb. Substituting in Equation 5.6 yields $P = 0.22$ for the proportion of peanuts. That is, 22% of the quantity must come out equal to the 200 lb of peanuts we have on hand. If we let N stand for the number of pounds of cashews we need (we can't call this variable C because C has already been used in the problem) then the total weight will be $200 + N$. Hence,

$$200 = 0.22(200 + N),$$

so

$$\frac{200}{0.22} = 200 + N,$$

and

$$N = \frac{200}{0.22} - 200 = 726 \text{ pounds of cashews.}$$

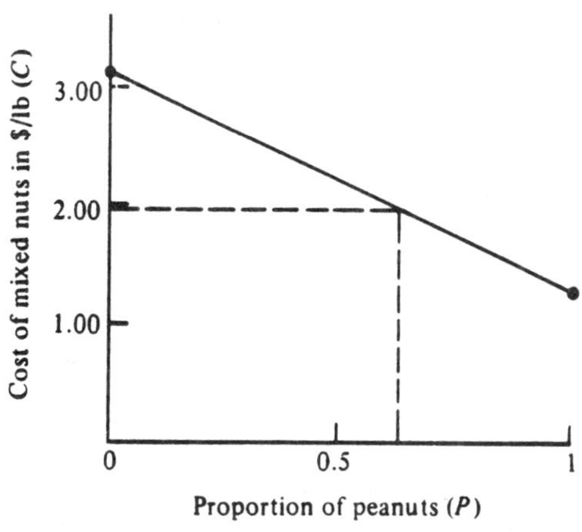

Figure 5.2 Finding the cost of a mixture of peanuts and cashews

EXAMPLE 3. DILUTING SULFURIC ACID

A chemist has on hand two solutions of sulfuric acid whose concentrations of the laboratory's standard solution are 7% and 60%. How much of each should she use to prepare 15 milliliters of a 48% solution?

As usual, we let the independent variable be the proportion P of one solution, say the weaker, in the mixture. When we have found the correct proportion we will then be able to compute the amounts asked for. The units of P are liters (or any convenient units of volume) of weak solution per liter of final mixture. P varies from 0 (all strong solution) to 1 (all weak solution). The quantity that depends on P is the concentration C of sulfuric acid in the mixture. The potentially confusing part of this problem is that C, the dependent variable, is a percentage and hence also a proportion. When $P = 0$, $C = 0.60$, because the mixture is then all strong solution, which is 60% acid. When $P = 1$, $C = 0.07$. The relationship between C and P is given by

$$C = 0.07P + 0.60(1 - P).$$

We want to know the value of P for which C will be 0.48. If we substitute 0.48 for C and solve for P we get 0.23. Thus the chemist's solution should consist of 23% of the weaker (7%) solution and 77% of the stronger. To make up 15 mL she will mix $0.23 * 15 = 3.4$ mL of the 7% solution with $15 - 3.4 = 11.6$ mL of the 60% solution.

EXAMPLE 4. WEIGHTED AVERAGES

A farmer plants two types of wheat. Type C produces c bushels/acre, type D produces d bushels/acre. What is his average yield?

One way in which this problem differs from the preceding ones is in the presence of unknown constants c and d where earlier we had numbers to deal with. But we can still solve it.

If the farmer plants the same number of acres of each type, then the average yield will just be the average

$$\frac{c + d}{2} \; \frac{\text{bushels}}{\text{acre}}. \tag{5.7}$$

But if he plants more of one kind than the other the average yield Y will vary. It will depend on the proportion P of type C wheat. In general,

$$Y = cP + d(1 - P). \tag{5.8}$$

Equation 5.7 is the special case of Equation 5.8 in which $P = 1/2$. If, for example, he plants 1300 acres of type C wheat and 2300 acres of type D wheat, then the proportion of his acreage devoted to type C wheat is

$$P = \frac{1300}{1300 + 2300} = 0.36.$$

(*Warning*: P is *not* 1300/2300.) Then

$$Y = 0.36c + 0.64d. \tag{5.9}$$

Notice that we could have found the same result by calculating

$$\text{total crop} = 1300c + 2300d \text{ (bushels)},$$

and then finding the average yield Y:

$$Y = \frac{\text{total crop}}{\text{number of acres}} = \frac{1300c + 2300d}{1300 + 2300}. \tag{5.10}$$

You should convince yourself that Equations 5.9 and 5.10 are equivalent.

PROBLEMS

*1. A farm cooperative makes a profit of 3.2 cents per can of economy-grade and 5.8 cents per can of fancy-grade applesauce. The deluxe cans account for 29% of their sales. What is the overall profit rate?

2. The state Department of Environmental Quality allows small energy facilities to burn oil whose sulfur content does not exceed 2.2%. Clean oil contains 1.2% sulfur; dirty oil contains 4.8% sulfur. How should those two fuels be mixed to meet the DEQ requirements?

*3. How many grams of solder with 60% tin must be mixed with 40% tin solder to produce 500 g of solder containing 55% tin?

4. Work Example 2 (peanuts and cashews) using Q, the proportion of cashews, as the independent variable.

*5. You have $1300 to invest in a combination of high-risk securities that pay 15.4% interest annually and low-risk securities that pay 9.1%. You must earn $140 per year to stay in school. How should you invest your money to do so at minimum risk?

6. You are in charge of investing $178 000 for a trust fund. You can buy Treasury Certificates, which return 11.27% annual interest, or put money in an ordinary savings account, which pays 5.65%.
   (a) How should you mix your investments in order to achieve an annual return of 8.25%?
   (b) Suppose that you must keep at least 40% of the money in savings, because certificates can't be cashed at a moment's notice, and that certificates can be bought only in multiples of $10 000. How should you invest the money so as to maximize the rate of return? At what rate will your mixed investment earn interest?

7. Supergreen fertilizer makes the grass grow 7 cm/month. Trueturf leads to a growth rate of 3.1 cm/month (but it's better for the roots). You use a mixture that is 55% Trueturf. How fast should this mixture make the grass grow?

*8. Your hardware and garden supply store sells two kinds of grass seed. Sturdyturf is 70% bluegrass and 30% ryegrass. Quickgrow is 20% bluegrass and 80% ryegrass. Your lawn, which is partly shaded, needs a mixture that is half bluegrass and half ryegrass.
   (a) What proportion of your mixture should be Sturdyturf and what proportion Quickgrow?
   (b) If bluegrass seed costs $1.80/lb and ryegrass seed costs $1.20/lb, find the cost per pound of Sturdyturf and of your mixture.

9. Dutch tulip bulbs cost $52.50 per hundred, domestic bulbs only $38.00 per hundred. The director of the Public Garden needs 1600 bulbs for fall planting and has budgeted $700 for their purchase. She wishes to buy as many of the better Dutch bulbs as possible. How many of each kind should she order?

10. Tap water is 50°F; water from the heater is 180°F.
    (a) How should they be mixed to give water at 145°F to do the laundry?
    (b) How much hot water (180°F) will you need to make 24 gallons of laundry water?
    (c) Suppose it costs 1/20 of a cent per gallon per degree to heat water. Compute the cost of heating 24 gallons of tap water in each of these ways, and compare the results: (1) Heat all the water from 50°F to 145°F. (2) Heat some of the water from 50°F to 180°F, then mix with tap water to reach the final temperature of 145°F.

11. An old 8-cylinder car greedily guzzles gas at the rate of 0.082

gal/mile but a compact car cruises comfortably at 0.037 gal/mile. The EPA would like the fuel consumption of the 100 million cars in the United States to average 0.053 gal/mile. Suppose, for a first approximation, that every car is either a gas guzzler or a compact.

(a) To achieve the EPA goal, what proportion of the cars must be compacts?

(b) Suppose there are 80 million gas guzzlers now. How many must be replaced by compacts to reach the goal?

(c) How many gallons of gas does a gas guzzler guzzle going 10 000 miles? How many does a compact use? If every car travels 10 000 miles each year, how many gallons of gas are used annually?

(d) How many gallons will be saved annually when the EPA goal is reached?

12. You wish to paint your room Sky Blue. The paint chip tells you that Sky Blue paint contains 5.4 grams of ozaloid blue pigment per liter. Unfortunately, Highland Hardware carries only Baby Blue (with 4.3 g of pigment/L) and Royal Blue (with 7.4 g/L).

(a) In what proportions should they be mixed to make Sky Blue?

(b) A 4-liter can of paint covers 18.5 square meters. If the room you wish to paint is 3.5 m wide, 4.3 m long, and 2.7 m high, how much paint will you need?

(c) How many 1- or 4-liter cans of each color, Baby Blue and Royal Blue, will you need?

*13. Two mixtures of water and alcohol are to be combined. Bottle #1 contains m parts alcohol per part mixture, bottle #2 contains M parts alcohol per part mixture. What is the concentration C of alcohol in a combination in which the fraction coming from bottle #1 is F? Solve this equation for F in terms of C, m, and M.

Now bottle #1 contains vermouth, $m = 0.16$ (32 proof), and the gin in bottle #2 is 90 proof ($M = 0.45$). A martini contains 1 part vermouth to 2 parts gin ($F = 1/3$, not $1/2$). What is the percentage of alcohol in the martini? The more gin, the *drier* the martini. What about a dry martini, which is 8 parts gin to 1 part vermouth? What's the driest possible martini? Does this problem count as applied algebra?

14. Redo Problem 38(b) of Chapter 4 using the techniques developed in this chapter.

*15. Milk has a density in the range of 1.023–1.035 kg/L. A dishonest dairy farmer has good cows that produce milk whose density is at the high end of the scale. How should he dilute the milk with water (density 1.000 kg/L) to bring the density of the mixture down to the low end of the scale?

*16. A private college charges $5700 tuition. Scholarship students receive free tuition and an allowance of $1250 for expenses. The college must collect an average of $4000 per student to meet expenses. What mix of paying and scholarship students will enable them to meet that goal?

17. Fifty-eight percent of a mixture is a 12% acid solution; the rest is an acid solution of unknown strength. If the mixture is 17% acid, find the concentration of the other component.

18. To entice customers, a department store advertises black-and-white television sets on sale at a price that causes them a loss of $18 per set. When the customers come in, the store hopes they will buy color instead, on which the profit is $81 per set. What proportion of the sales must be color sets in order for the store to show a net profit?

*19. Suppose macadamia nuts cost $7.28 per pound. Use the prices for peanuts and cashews in Example 2 to calculate the cost of a mixture of 6% macadamia nuts, 23% cashews, and the remainder peanuts.

20. Gorp is a mixture of good old raisins and peanuts—plus chocolate chips. A pound of each of these three ingredients costs, respectively, $1.63, $1.37, and $4.17. In order to taste right, there should be one and one half times as many raisins as chocolate chips. Find the proportions of the various ingredients in Gorp that costs $2.00 per pound.

CHAPTER 6

Simultaneous Equations

> But suppose I have a piece of meadow that is three times as large as the field of my neighbor Dagantakala, but he buyeth a yoke of land in addition and now mine is only twice as large as his. How many yoke have both fields?
>
> – Thomas Mann, *Joseph and His Brothers*

Comparisons / Effects with multiple causes

EXAMPLE 1. TREZH AND SIVA

The Trezh Car Rental Company charges an initial \$32 and then 27 cents per mile. Siva, Inc., charges \$28, and then 30 cents per mile. Compare their rates for trips of various lengths. In particular, for how long a trip would the cost be the same with each company?

The variables in this problem are C, the cost in dollars of renting a car for a trip, which depends on L, the length in miles of the trip. How C depends on L depends on which company we are renting from. There will be two equations. Each is linear. In fact, we have been given the intercepts and slopes for both. Trezh charges

$$C = 32 + 0.27L \qquad (6.1)$$

for a trip of L miles. The same rental from Siva costs

$$C = 28 + 0.30L. \qquad (6.2)$$

The graphs for these equations are drawn in Figure 6.1, which should look familiar if you worked Problem 15 in Chapter 1. That graph shows

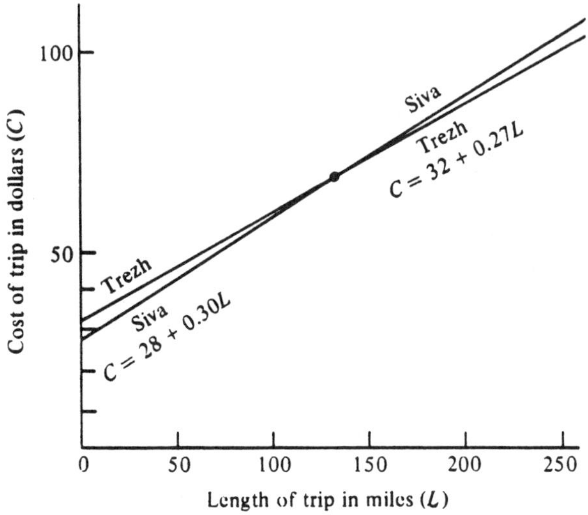

Figure 6.1 Car rental costs

that Trezh cars are cheaper for long trips and Siva cars for short ones. The changeover point is between 100 and 150 miles. To find it exactly we must solve Equations 6.1 and 6.2 *simultaneously.* We want to find the particular values for C and L that make both equations true. The easiest way to do so is to equate the right halves of each, converting the two equations in the two unknowns L and C into the single equation

$$32 + 0.27L = 28 + 0.30L.$$

This involves only L, and so we can solve it in the usual way. Collecting the L's on the left side gives

$$0.30L - 0.27L = 32 - 28,$$

so

$$0.03L = 4$$

and

$$L = \frac{4}{0.03} = 133 \text{ miles.}$$

To find the common cost for this long a trip, substitute $L = 133$ in either of Equations 6.1 or 6.2 to conclude that $C = \$68$.

EXAMPLE 2. BERKELEY PLASTICS REVISITED

Berkeley Plastics' monthly costs depend linearly on the number of panels they manufacture in a month. If their July costs were \$8044 to make 12 panels, while it cost them \$22 052 to make 46 in August, find the equation showing how to calculate their costs in terms of the number of panels made.

Let C be the monthly cost in dollars when N panels are manufactured. Then

$$C = \text{intercept} + \text{slope} * N. \tag{6.3}$$

Our job is to find the intercept, which is BP's overhead, and the slope, which is their cost of materials in dollars per panel.

We solved this problem in Example 1 of Chapter 2. We are about to solve it again, in a new way. If we substitute the data for July and August in turn in Equation 6.3, we obtain the two equations

$$8044 = \text{intercept} + \text{slope} * 12 \tag{6.4}$$

and

$$22\,052 = \text{intercept} + \text{slope} * 46. \tag{6.5}$$

Now we treat Equations 6.4 and 6.5 as simultaneous equations in the two unknowns *intercept* and *slope*, and find values that make both equations true. As in Example 1, we start by converting the pair of equations into a single equation involving just one of the unknowns. In this problem, such an equation will result if we subtract one of the two from the other, so that the intercept disappears.

To keep the numbers positive, hence less confusing, we shall subtract Equation 6.4 from Equation 6.5. The result is

$$\begin{aligned} 22\,052 - 8044 &= (\text{slope} * 46) - (\text{slope} * 12) \\ &= \text{slope} * (46 - 12) \end{aligned}$$

so

$$\text{slope} = \frac{22\,052 - 8044}{46 - 12} = 412 \, \frac{\$}{\text{panel}}.$$

This is, of course, the traditional way in which we find the slope. To recover the intercept, substitute the value 412 \$/panel for the slope in either of Equations 6.4 or 6.5. For example, Equation 6.4 yields

$$8044 = \text{intercept} + 412 * 12$$

so

$$\text{intercept} = 8044 - 412 * 12 = \$3100.$$

To check this result, test it in Equation 6.5.

EXAMPLE 3. THE HAMBURGER BUSINESS

A hamburger stand should live in a densely populated area on a heavily traveled street. Experience shows that the weekly gross sales G (in thousands of dollars) of such a stand can be found using the formula

$$G = kP + bC, \tag{6.6}$$

where P is the population in thousands living within three miles of the stand, C is the number of thousands of cars which pass the stand each day, and k and b are constants. Find k and b if G = \$4900 when $P = 10.1$ and $C = 7.2$, while G = \$6400 when $P = 15.3$ and $C = 4.4$.

We start by substituting the data for each of the two businesses in Equation 6.6. This tells us that

$$4900 = 10.1k + 7.2b \tag{6.7}$$

and

$$6400 = 15.3k + 4.4b. \tag{6.8}$$

We must use these two equations to find the values of the two unknowns k and b. This time it is harder to eliminate one of them and find a single equation involving the other. The following systematic procedure works best: If we multiply each side of Equation 6.7 by 15.3 (the coefficient of k in Equation 6.8), and each side of Equation 6.8 by 10.1 (the coefficient of k in Equation 6.7), then the resulting pair of equations is

$$15.3 * 4900 = 15.3 * 10.1k + 15.3 * 7.2b$$

and

$$10.1 * 6400 = 10.1 * 15.3k + 10.1 * 4.4b.$$

Now the coefficient of k is the same in each of these equations, so if we subtract the second from the first, the new equation will involve only b:

$$15.3 * 4900 - 10.1 * 6400 = (15.3 * 7.2 - 10.1 * 4.4)b$$

so

$$b = \frac{15.3 * 4900 - 10.1 * 6400}{15.3 * 7.2 - 10.1 * 4.4}$$

$$= 157 \frac{\$}{\text{thousand cars}}$$

$$= 0.157 \frac{\$}{\text{car}}.$$

Thus each car that passes the stand each day contributes an average of about 16 cents to the weekly gross.

We can now find k by substituting the known value of b in either Equation 6.7 or 6.8, or by following the same kind of strategy to eliminate b. The latter would require multiplying Equation 6.7 by 4.4 and Equation 6.8 by 7.2, giving

$$4.4 * 4900 = 4.4 * 10.1k + 4.4 * 7.2b$$

and

$$7.2 * 6400 = 7.2 * 15.3k + 7.2 * 4.4b.$$

Then subtracting the first from the second we get

$$7.2 * 6400 - 4.4 * 4900 = (7.2 * 15.3 - 4.4 * 10.1)k.$$

Thus

$$k = \frac{7.2 * 6400 - 4.4 * 4900}{7.2 * 15.3 - 4.4 * 10.1}$$

$$= 373 \frac{\$}{\text{thousand people}}$$

$$= 0.373 \frac{\$}{\text{person}}.$$

On the average, each person in the neighborhood contributes about 37 cents to the weekly gross.

PROBLEMS

*1. In 1960 the average American drank 33 gallons of milk and 14 gallons of soft drinks. In 1979 the figures were 28 and 38 gallons, respectively. Assuming that the average annual consumptions varied linearly, graph both lines and determine when milk and soft-drink consumption were equal.

2. A salesman's earnings depend on the volume of business he can do. He is offered a choice by his employer: a weekly wage of \$250 plus a commission of 1% of the dollar value of his sales, or \$100 in wages plus a 2.3% commission. Describe which plan is better for him for various sales volumes.

3. Solve Problem 6 of Chapter 2 by the method illustrated in Example 2 of this chapter.

4. When do the Fahrenheit and Celsius temperature scales give the same reading? See Problem 10(d) of Chapter 2.

5. Solve Problem 9 of Chapter 5 by using for two unknowns the number of domestic and imported bulbs and writing two equations involving these unknowns, one saying that the total number of bulbs is 1600, the other that the total cost is \$700.

*6. A movie theater sells 595 adult tickets and 680 children's tickets and receives \$3183.25. For the next performance, at the same prices, it sells 1025 adult and 130 children's tickets and receives \$4025.75. Find the price of each kind of ticket.

7. Supercold refrigerators cost \$1000 and depreciate at a rate of \$30 per year. Iceking brand costs \$1200 and depreciates \$40 per year. Draw the depreciation graphs on the same set of coordinate axes. When do the refrigerators have the same value?

*8. The duration T of a subway trip in minutes depends on its length L in miles and on the number S of stops according to the formula

$$T = cS + dL,$$

where c and d are constants.

(a) What are the units of c and d?

(b) A 15-mile trip with 9 stops takes 40 minutes; a 4-mile trip with 3 stops takes 12 minutes. Find c and d.

9. To manage Mr. Gaffney's retirement account, his bank charges either \$3 per month plus one half of one percent of the account

per year, or 0.6% of the account per year, whichever is less. For what size retirement account do the methods yield the same fee?

*10. The cost C of manufacturing N calculators is given by

$$C = D + mN,$$

where D is the design and setup cost and m is the cost in materials and labor required to make each calculator. If calculators sell for S dollars each, then the income I that results from the sale of N calculators is

$$I = SN.$$

The breakeven point is the number of calculators that must be sold in order to have $I = C$. Find it, in terms of D, m, and S. (Compare with Problem 18 of Chapter 1.)

11. You are about to go into business renting cars and wish to set your rates so as to be cheaper than both Trezh and Siva precisely for all trips between 50 and 200 miles. Use the data in Example 1 to figure out what your initial fee and mileage charge should be.

12. At what monthly production level do Wing Aero and Berkeley Plastics incur essentially the same total cost? (See Example 1 of Chapter 2.)

13. Solve the problem asked of Joseph in the quotation from Thomas Mann that opens this chapter. The quotation continues:

> "Together?" asked Joseph and reckoned.
> "No, each one."
> "Hast thou a neighbour named Dagantakala?"
> "I only call so the owner of the second field in my sum."
> "I see and understand. Dagantakala–that must be a man from the country of Pelesheth to judge by the name, from the land of the Philistines, whither we seem to be going down according to the decree of thy mind. There is no such person; but he is named Dagantakala and he tills in contentment his little ploughland, now increased to three yoke, incapable of envy of my lord and his six yoke, since after all he hath increased from two to three yoke and besides, because he doth not exist at all nor yet the ploughland which all together maketh up nine yoke–that is the joke of it. There is only my lord and his busy brain."
> The old man blinked uncertainly, for he did not quite see that Joseph had already solved the problem.
> "Well?" asked he. . . . "Ah, yes, yes. Thou hast said already and I scarce marked it, so hast thou woven and fabled it into thy prattle that I almost

failed to hear it. It is right, six, two and three, those are the figures. They were concealed and hidden–how hast thou so quickly brought them out while prattling?"

"One must fix the unknown quantity clearly in one's eye, then the concealments fall away and it becometh known."

CHAPTER 7

Problems Using Linear Techniques

Three minutes' thought would suffice to find this out, but thought is irksome and three minutes is a long time.

– A. E. Housman, *Juvenalis Saturae*

1. Overtime.
   Carol Williams has a summer job as a laboratory technician. She is paid \$3.80/hour for the first forty hours she works in a week, and then time and a half (\$5.70/hour) for overtime.
   (a) Write equation(s) and draw a graph showing how her earnings in a week depend on how many hours she works.
   (b) Why is this graph called *piecewise linear*?
   (c) How much does she earn when she works 47 hours?
   (d) How long would she have to work to earn \$175?
   (e) Write equation(s) and draw a graph showing how the number of hours she must work depends on what she hopes to earn.

2. A piecewise linear tax structure.
   A simplified model of a graduated income tax structure like ours says that one pays no tax at all on income up to \$1500. For income between \$1500 and \$4500 the tax rate is 9%. For income between \$4500 and \$15 000 the tax rate is 14%. On the next \$35 000 one pays 28%. On all the rest of one's income the tax is 50%.
   (a) Draw two graphs, one showing how the amount of tax paid depends on income, one showing how income depends on tax paid.
   (b) On what income is the overall, net tax rate 25%?

3. Step functions.
At the time of writing it costs 20 cents for each ounce or part thereof to send first class mail.
   (a) How much does it cost to mail letters weighing 1.3, 1.8, 2.0, and 2.3 ounces?
   (b) Draw a graph showing how the postage depends on the weight of the mail being sent. Why is this graph called a *step function*?

4. A weighted average.
The Arlington Five Cent Savings Bank has a mortgage portfolio of \$111 million, earning from a low of 4.5% to today's 14.5%. A rough breakdown follows. Since more precise information is not available, make, and justify, the approximations needed to answer the following questions.
   (a) One of the columns in Table 7.1 is redundant and can be used to check another. Do that.
   (b) The president of the bank says the entire mortgage portfolio is earning only 9.51%. Confirm her assertion.
   (c) If the money now outstanding at 4.5–6% were repaid and reinvested, what reinvestment rate would be necessary to bring the Arlington Five's overall rate of return up to 10%?

Table 7.1 Arlington Five–mortgages outstanding

Mortgage percentage rate	Balance outstanding	Percentage of mortgage loans
4.5 - 6	\$ 9 095 739	8.13
6.25- 8	4 282 296	3.83
8.25–10	58 494 507	57.27
10.25–12	19 805 145	17.70
12.25–14	13 758 184	12.30
over 14	6 458 484	5.77

5. Fuel Economy.
A General Motors Corporation advertisement in the December, 1978 *Scientific American* states that the company has discovered that for any particular car, the fuel consumption rate F, in liters per kilometer, for city driving depends linearly on the rate of travel A, in hours per kilometer (including idling in traffic). Let us explore the consequences of this discovery. Start with the estimates that for a compact car, $F = 0.2$ L/km when $A = 0.05$ hr/km, and $F = 0.15$ L/km when $A = 0.025$ hr/km.
   (a) What are these speeds in km/hr and in mi/hr?

(b) Find the equation for and sketch the graph of the dependence of F on A. What are the units of the slope and the intercept?

(c) Find F when $A = 0.03$ hr/km, and A when $F = 0.18$ L/km.

(d) General Motors says that the slope of the line in (b) depends on the rate at which the car burns gas when it is idling, and that the intercept depends on the weight of the car. Explain why this interpretation is reasonable.

(e) Suppose that the slope and intercept of the line in (b) are in fact each directly proportional to the weight of the car. (Why is that a reasonable assumption?) Find out how much a typical compact car and a typical full-size car weigh, and figure out the slope and intercept of the line in (b) for the full-size car. Draw the graph showing how F depends on A for that car, on the same set of coordinate axes you used before.

(f) Demonstrate that for a fixed A, F is proportional to the weight of the car.

(g) What policies or laws or purchases can you think of for yourself, or for GM or for the government, which would reduce F by: (1) decreasing A? (That was the point of the ad.) (2) decreasing the slope? (3) decreasing the intercept?

6. Recognizing qualitative behavior.
The labels have been left off the six graphs in Figure 7.1. Please choose the correct ones from the following list. Name the variables, with their units. Where possible, put reasonable scales on the axes. Notice that there is one more graph than caption. Make up a caption to go with that graph. Identify which graphs are linear, and which represent direct proportion.

(a) How the average daily temperature in Boston depends on the day of the year.

(b) How the number of cookies in the cookie jar depends on the length of time since the cookies left the oven.

(c) How the cost (in \$/mile) of driving a car depends on the age of the car.

(d) What you pay for heating oil depends on how much you buy.

(e) How the value of a house changes with time.

7. Musings from the market.
At the supermarket checkout counter the clerk needs 8 seconds to ring up and bag each item, and 2 minutes per customer to ring up the total sale and make change.

(a) If there are 6 people in the express line, each of whom has 7

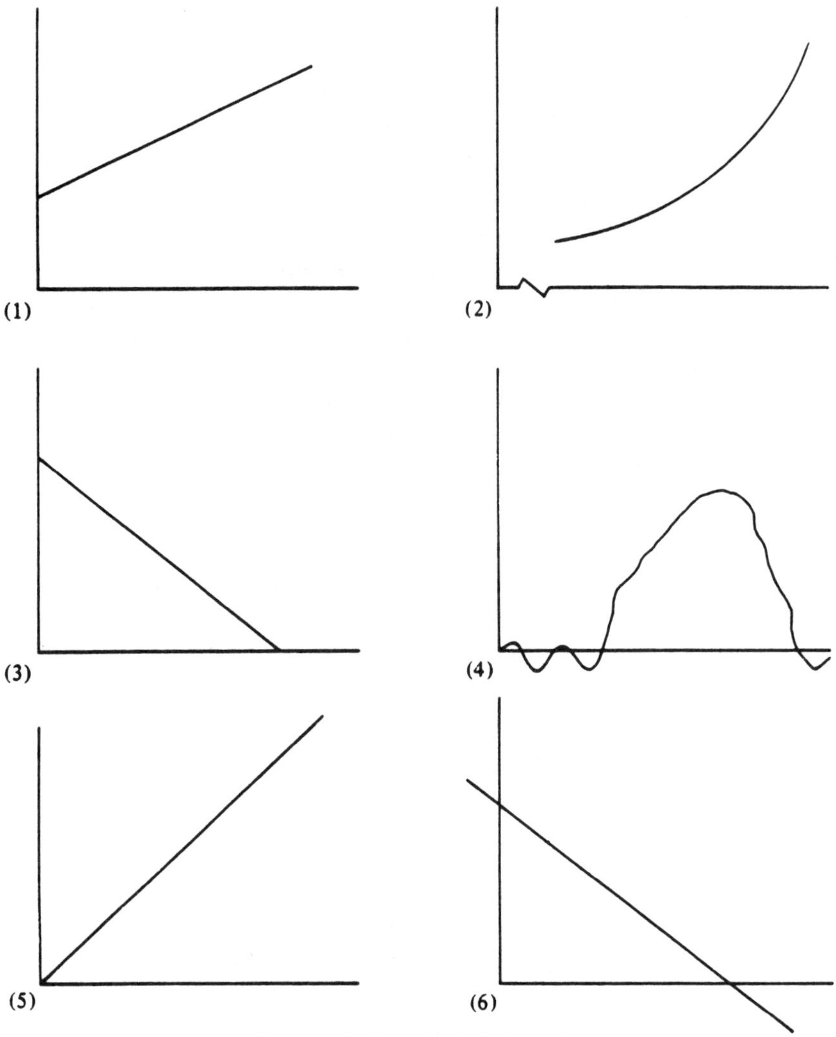

Figure 7.1 Six unlabeled graphs

items, and 2 in the regular line, each of whom has 25 items, and you arrive with 8 items, which line should you stand in?

(b) If the store manager anticipates 130 customers per hour, each of whom buys, on the average, 15 items, how many checkout lines will she have to open?

8. For February 14.
The *Boston Globe* advertises a special rate for Valentine's Day personal ads:

3 lines for \$10.20,
4 lines for \$13.60,
5 lines for \$17.00.

(a) Does the cost depend linearly on the number of lines?

(b) What would you expect to pay for a seven-and-one-half-line message?

(c) How did Valentine's Day get its name?

9. Shipping charges.
Compare two models for computing shipping charges in terms of the weight W of the item and the distance D it is sent:

$$C = F + aW + bD$$

and

$$C = F' + kWD.$$

F and F' are fixed flat fees; a, b, and k are constants.

(a) What are the units of the constants F, F', a, b, and k, in terms of appropriate units for C, W, and D?

(b) Suppose both models have the same fixed fee, and charge the same amount for shipping a 10-pound package 10 miles. Which is likely to be cheaper for shipping a 1-pound package 100 miles?

10. Octane ratings.
The octane rating of gasoline was originally defined as the percentage of *isooctane* it contained. Thus octane ratings could vary from 0 to 100. Unfortunately, isooctane was both expensive and necessary to run airplane engines, so that during World War II a search was begun to find ways to make cheaper ingredients burn like 100-octane gas. It turned out that by adding a compound of lead one could make "100-octane" gas that was not 100% isooctane. In fact, by extending the scale linearly, measuring burning quality rather than percentage of isooctane, it is possible to have gasolines with octane ratings greater than 100.

(a) A refinery makes high-octane gas, rated at 132, and low-octane gas, rated at 73.4. How should these be mixed to produce 400 000 barrels of 90-octane gas?

(b) How should the refinery's low-octane gas be mixed with pure isooctane to yield 90-octane gas?

(c) The low-octane gas costs \$1.24 per gallon; the high-octane \$1.93 per gallon. Pure isooctane costs \$1.82/gallon. Compare the costs of making 90-octane gasoline in each of the ways described in parts (a) and (b).

11. Mixed nuts.
It costs Royal Nut Farm \$17.70 to prepare and fill a 4-pound box of Royalmix with almonds. A 6.5-pound box costs them \$26.25. Preparing and filling the 5-pound Economy box (no almonds) costs \$15.75. Suppose that the preparation time and the cost of the empty box are essentially constant. Find the cost of buying, preparing, and filling an empty box; the cost of almonds; and the cost of the rest of the mixture.

12. Dieting.
You must eat about 2500 Calories per day just to maintain your current weight and generate the energy you need to live and work. In order to lose weight you should eat less. Each pound of body fat is equivalent to 3300 Calories.

(a) To lose L pounds, you can cut C Calories per day from your maintenance diet for D days. For fixed L, write an equation showing how C and D are related.

(b) Write an equation showing how your allowed daily intake I depends on the amount of weight you want to lose and the length of time you plan to diet.

(c) Suppose you wish to lose 10 pounds. Draw a graph showing how your daily Calorie intake depends on the length of time you plan to diet.

13. Resistance.
The resistance R (in ohms) of a wire is directly proportional to its length L (in cm) and inversely proportional to the square of its radius r (in cm). If a 10-m wire of radius 1 mm has a resistance of 10 ohms, find the resistance of a 100-foot wire of diameter 0.1 inch.

14. Depreciation.
Figure 7.2 shows three depreciation curves. Match them to the following descriptions. In each case describe the depreciation qualitatively and explain why your choice is appropriate. (Ignore the effects of inflation.)

(a) The value of a commercial refrigerator.

(b) The value of a car.

(c) The value of the oil remaining in a small oil well as it is being pumped out.

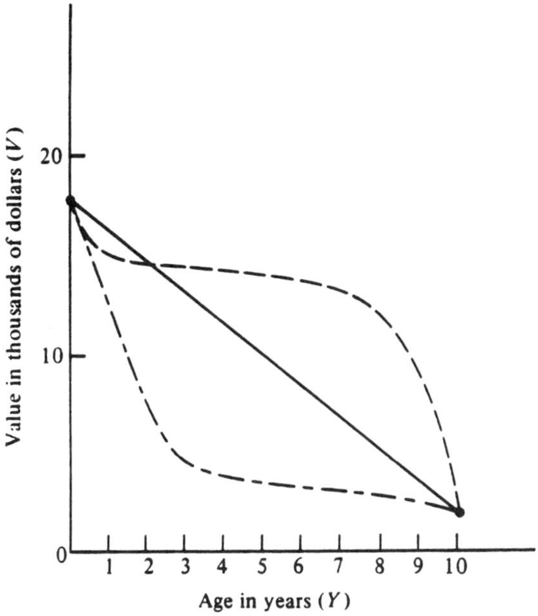

Figure 7.2 Three depreciation curves

15. Square roots.

Your calculator has a square-root key. In this problem we will see how you could do without it. We shall see that if A is an approximation to $\sqrt{T}$ then

$$A' = \frac{1}{2}\left(A + \frac{T}{A}\right)$$

is a better one.

(a) Suppose $T = 177$. Find A' when $A = 12$. Store A' in the calculator's memory.

(b) Square both A and A'. Which is a better approximation to $\sqrt{177}$?

(c) T is still 177. Repeat steps (a) and (b), using for A the value of A' you just found and stored in the memory. That produces a new A.

(d) Repeat step (c) until nothing seems to change. How many repetitions does that take?

(e) Square your last answer. Comment.

Part II

Exponential Models

CHAPTER 8

Large Numbers

Yet the number of the children of Israel shall be as the sand of the sea, which cannot be measured nor numbered.

– *Hosea* II, 1

Handling very large and very small numbers / Scientific notation / Estimation / Review of the rules for manipulating exponents

EXAMPLE 1. THE SPEED OF LIGHT

Light travels about 186 000 miles per second. Find the speed of light in km/year and in cm/year.

These are familiar unit conversions. To change 186 000 mi/sec to km/yr we multiply

$$186\,000\,\frac{\cancel{\text{mi}}}{\cancel{\text{sec}}} * 1.609\,\frac{\text{km}}{\cancel{\text{mi}}} * 60\,\frac{\cancel{\text{sec}}}{\cancel{\text{min}}} * 60\,\frac{\cancel{\text{min}}}{\cancel{\text{hr}}} * 24\,\frac{\cancel{\text{hr}}}{\cancel{\text{day}}} * 365.25\,\frac{\cancel{\text{days}}}{\text{yr}}$$

$$\begin{aligned} &= 9\,444\,370\,000\,000 \text{ km/yr} \\ &\sim 9\,440\,000\,000\,000 \text{ km/yr.} \end{aligned} \tag{8.1}$$

We have rounded the answer to three significant digits because that is all the data contain. There is a conventional method for keeping track of the number of digits (including zeros) in the last answer in Equation 8.1: Place the decimal point just after the first significant digit (9), then tell the reader how far to move it back to its proper place. Thus:

9 440 000 000 000 = 9.44 (move decimal point 12 places to the right).

Next, recall that multiplying by 10 is equivalent to moving the decimal point one place to the right; to move it 12 places to the right, multiply by 10 12 times. That is, multiply by 10 * 10 * 10 * 10 * 10 * 10 * 10 * 10 * 10 * 10 * 10 * 10, which is 10^{12}. (*Warning*: Multiplying by 10 12 times is *not* multiplying by 120!) Thus, the answer to the first calculation may be expressed as

$$9.44 * 10^{12} \text{ km/yr}. \tag{8.2}$$

This is read: "Nine point four four times ten to the twelfth kilometers per year." A number written in terms of ten to some power with the decimal point so placed is said to be written in *scientific* or *exponential notation*.

A good calculator displays the number in expression 8.2 in another way, since it is unable to raise exponents half a line. If you use yours to solve the multiplication problem in Equation 8.1 it will display the answer

$$9.444\,37 \qquad 12 \tag{8.3}$$

Since it cannot show all 13 digits in Equation 8.1, it shifts automatically to scientific notation, placing the power to which 10 is raised at the right of its display. Some calculators can be instructed to show shorter numbers, which would fit within the display, in scientific notation. Can yours? (*Warning*: The display in Equation 8.3 is read: "9.444 37 times 10 to the 12th," *not* "9.444 37 to the 12th.")

We have one more conversion to make, that of $9.44 * 10^{12}$ km/yr to cm/yr. Since there are 10^3 m/km and 10^2 cm/m the answer is

$$9.44 * 10^{12} \frac{\cancel{\text{km}}}{\text{yr}} * 10^3 \frac{\cancel{\text{m}}}{\cancel{\text{km}}} * 10^2 \frac{\text{cm}}{\cancel{\text{m}}} = 9.44 * 10^{17} \frac{\text{cm}}{\text{yr}}. \tag{8.4}$$

Here we have used the important principle that when you *multiply* several powers of the same number you *add* the exponents:

$$10^{12} * 10^3 * 10^2 = 10^{(12+3+2)} = 10^{17}. \tag{8.5}$$

Verify Equation 8.4 with your calculator, entering 10^3 and 10^2 as 1000 and 100.

EXAMPLE 2. THE SPEED OF LIGHT, CONTINUED

Radio waves and electrical signals travel at the same speed as light. How long does it take a radio signal to travel to the sun? How long does it take an electrical impulse to cross a 1-cm wide silicon chip in your calculator?

To answer the first question we must know how far away the sun is. A good dictionary or encyclopedia (or a visit to the reference section of the library) will yield the figure 93 million, or 93 000 000, or $9.3 * 10^7$ miles. Thus the time it takes the radio signal to reach the sun is

$$\begin{aligned}\frac{\text{distance (miles)}}{\text{rate (mi/sec)}} &= \frac{9.3 * 10^7}{1.86 * 10^5} \\ &= \frac{9.3}{1.86} * 10^2 \\ &= 5.0 * 10^2 \text{ sec.}\end{aligned} \tag{8.6}$$

This calculation illustrates the principle that when you *divide* one power of a number by another power of the same number you *subtract* the exponents:

$$\frac{10^7}{10^5} = 10^{(7-5)} = 10^2.$$

The answer to the first question is 500 seconds, or about 8⅓ minutes. We can use this fact to try to grasp the enormity of the number $9.44 * 10^{12}$ km, which we found in Example 1 is the distance light travels in a year. That is a unit of distance called a *light-year*. Do not be confused by the fact that the unit of distance *light-year* mentions time.

We often measure distances in terms of time, as for instance: "I live 45 minutes from work," or "a kilometer is a ten-minute brisk walk." If you should read that something happened *light-years ago*, you will know the writer is confused. Something can happen light-years *away*, but not light-years *ago*.

There are objects that are light-years away. The sun is the nearest star. The second nearest, Alpha Centauri, is 4.4 light-years, or

$$4.4 * 9.44 * 10^{12} = 41.536 * 10^{12} \sim 4.2 * 10^{13} \text{ km}$$

away. If it were to explode today we would not know about it for more than four years.

Light covers large distances rapidly. The second question in this example asks how long it takes to travel a short distance. We suspect that the answer will be a very small number of seconds. To find it, we first convert the speed of light in miles/sec to cm/sec.

$$186\,000\ \frac{\cancel{\text{mi}}}{\text{sec}} * 1.609\ \frac{\cancel{\text{km}}}{\cancel{\text{mi}}} * 10^5\ \frac{\text{cm}}{\cancel{\text{km}}}$$

$$= 2.9927 * 10^{10}\ \frac{\text{cm}}{\text{sec}} \tag{8.7}$$

$$\sim 3 * 10^{10}\ \frac{\text{cm}}{\text{sec}}.$$

That rounds nicely to give the speed of light as roughly $3 * 10^{10}$, or 30 billion, cm/sec. To find the time it takes to travel 1 cm, we need only find the reciprocal of the speed. The answer will be very small—about one thirty-billionth of a second. Applying the [1/*x*] key on the calculator to the answer in Equation 8.7, we get

$$\frac{1}{2.9927 * 10^{10}\ \frac{\text{cm}}{\text{sec}}} = 3.3414 * 10^{-11}\ \frac{\text{sec}}{\text{cm}} \tag{8.8}$$

$$= 0.000\,000\,000\,033\,414\ \frac{\text{sec}}{\text{cm}}.$$

The exponent −11 in Equation 8.8 means that the decimal point must move 11 places to the left. This illustrates the principle that *multiplying* by 10^{-11} is the same as *dividing* by 10^{11}.

EXAMPLE 3. SCIENTIFIC NOTATION

In this example we explain how to write numbers in scientific notation, with and without the calculator.

To write 237.6 in scientific notation, place the decimal point between the 2 and the 3 and indicate how many times you must multiply by 10 to restore it to its proper place between the 7 and the 6:

$$237.6 = 2.376 * 10^2.$$

To write $237.6 * 10^9$ in scientific notation we must move the decimal point as above and then add the exponents of 10:

$$237.6 * 10^9 = 2.376 * 10^2 * 10^9 = 2.376 * 10^{11}.$$

Similarly,

$$237.6 * 10^{-53} = 2.376 * 10^2 * 10^{-53} = 2.376 * 10^{-51}.$$

To write 0.002 34 in scientific notation we must put the decimal point between the 2 and the 3. Restoring it to its proper place would then require three shifts to the left, so

$$0.002\,34 = 2.34 * 10^{-3}.$$

To write 6.2 in scientific notation we do not need to move the decimal point at all:

$$6.2 = 6.2 * 10^{0}.$$

Moving the decimal point 0 places is the same as multiplying by 1: nothing changes. That is why 10^0 *is* 1. The number 1 itself in scientific notation is written: $1 = 1 * 10^0$.

In elementary school you learned that 8, 8.0, and 8.0000 were all the same. That is true when you think of them purely as numbers, but false when they represent measurements. A scientist might report the result of a weighing as 8 g, as 8.0 g, or as 8.0000 g. The first report means "the number of grams is between 7.5 and 8.5." The second means "the number of grams is between 7.95 and 8.05." The zero following the decimal point in this case is significant; it indicates the accuracy of the measurement. Thus, 8.0000 g means that the number reported is between 7.99995 and 8.00005 g. Scientific notation makes it easy to show how many digits are significant. The speed of light, reported with 1, 2,. 3, 4, 5, and 6 significant digits, is $3 * 10^{10}$, $3.0 * 10^{10}$, $3.00 * 10^{10}$, $2.998 * 10^{10}$, $2.9979 * 10^{10}$, and $2.997\,93 * 10^{10}$ cm/sec. In particular, the fourth digit on the right in Equation 8.7 is wrong. That is not surprising, since the number we started with, 186 000 or $1.86 * 10^5$ miles/sec, had only three significant digits.

Here are two sample calculations in scientific notation. First we carry out the inversion in Equation 8.8 without using the calculator to keep track of the exponent.

$$\frac{1}{2.9927 * 10^{10} \frac{\text{cm}}{\text{sec}}} = \frac{1}{2.9927} * 10^{-10} \frac{\text{sec}}{\text{cm}}$$

$$= 0.33414 * 10^{-10} \frac{\text{sec}}{\text{cm}}$$

$$= 3.3414 * 10^{-11} \frac{\text{sec}}{\text{cm}}.$$

Here is one with more arithmetic.

$$\frac{(7.632 * 10^{21})(9.4 * 10^{-5})}{3.81 * 10^{3}}$$

$$= \frac{7.632 * 9.4}{3.81} * 10^{(21-5-3)}$$

$$= 18.8296 * 10^{13}$$

$$= 1.9 * 10^{14}.$$

We calculated with the exponents separately in the first line, then rounded the answer to 2 significant digits, since that is the accuracy of the least accurate of the numbers we started with. We moved the decimal point one place to express the answer in scientific notation.

To do the work in the second example entirely on the calculator, without treating the exponents separately, you must learn how to enter the number $7.632 * 10^{21}$. Use the [EE] key. EE means *enter exponent*. A sequence like

7.632 [EE] 21

will put $7.632 * 10^{21}$ into the machine. The display will show

7.632   21.

A *word of warning*: To make your calculator display 10^{21} you must enter

1 [EE] 21,

not

10 [EE] 21.

The latter means $10 * 10^{21}$, or 10^{22}.

Consult your instruction manual for help. On some calculators the *enter exponent* key is labeled [EEX] or [EXP]; on others, [EXP] means something entirely different. Learn how yours works.

EXAMPLE 4. THE SAND RECKONER

> There are some, King Gelon, who think that the number of the sand is infinite in multitude; and I mean by the sand not only that which exists about Syracuse and the rest of Sicily but also that which is found in every region whether inhabited or uninhabited. Again there are some who, without regarding it as infinite, yet think no number has been named which is great enough to exceed its multitude. And it is clear that they who hold this view, if they imagined a mass made up of sand in other respects as large as the mass of the earth, including in it all the seas and the hollows of the earth filled up to a height equal to that of the highest of the mountains, would be many times further still from recognising that any number could be expressed which exceeded the multitude of the sand so taken.

This quotation is from Archimedes' "The Sand Reckoner." We will follow him in spirit, and estimate the number of grains of sand it would take to make the entire earth of sand. We start more modestly by asking: How many grains of sand are there in a cubic centimeter? Once we estimate that number we need only multiply by the volume of the earth, measured in cubic centimeters. A square centimeter is about the area of a little fingernail. How many grains of sand will it carry? No exact answer is possible, for we might be talking about fine beach sand, or coarse builder's sand. But a fair rough estimate might be that 20 grains make a row 1 cm long. Then 20 rows of 20 grains, or $20 * 20 = 4 * 10^2$ grains, could rest on a little fingernail, and 20 layers of 400 grains would fill a box 1 cm on each edge. Thus we are led to guess that there are perhaps 8000, or

$$8 * 10^3 \text{ grains/cm}^3. \tag{8.9}$$

Next, let us compute the volume of the earth. The earth's radius is about 4000 miles, or about

$$4000 \text{ mi} * 1.6 \text{ km/mi} = 6400 \text{ km} = 6.4 * 10^3 \text{ km}.$$

Then we look up the formula for the volume of a sphere in terms of its radius, and find that

$$\text{volume} = \frac{4\pi}{3} \text{ radius}^3.$$

Thus the volume of the earth is about

$$\begin{aligned}\frac{4\pi}{3} * (6.4 * 10^3 \text{ km})^3 &= \frac{4\pi}{3} * 6.4^3 * (10^3)^3 \text{ km}^3 \\ &= 1098.07 * 10^9 \text{ km}^3 \qquad (8.10) \\ &\sim 1.1 * 10^{12} \text{ km}^3.\end{aligned}$$

(Use the value for π permanently stored in your calculator. If it has none, use $\pi \sim 3.14$).

In this calculation we used the fact that

$$(a * b)^3 = a^3 * b^3$$

(where $a = 6.4$ and $b = 10^3$). We also remembered to cube the units: a radius measured in kilometers yields a volume measured in cubic kilometers, km^3. Then we *multiplied* the exponents 3 and 3 to find the exponent 9, because

$$\begin{aligned}(10^3)^3 &= 10^3 * 10^3 * 10^3 \\ &= 10^{(3+3+3)} \\ &= 10^{3*3} = 10^9.\end{aligned}$$

This illustrates the principle that:

$$(10^a)^b = 10^{a*b}. \qquad (8.11)$$

Finally, in order to finish our estimate we must convert the volume in Equation 8.10, which is measured in km^3, to cm^3. There are 10^5 cm/km, and hence

$$\left(10^5 \frac{\text{cm}}{\text{km}}\right)^3 = 10^{15} \frac{\text{cm}^3}{\text{km}^3}.$$

We have used the principle in Equation 8.11 again. Finally, we find the number of grains by multiplying

$$\begin{aligned}8 * 10^3 \frac{\text{grains}}{\cancel{\text{cm}^3}} * 10^{15} \frac{\cancel{\text{cm}^3}}{\cancel{\text{km}^3}} * 1.1 * 10^{12} \cancel{\text{km}^3} \\ &= 8.8 * 10^{30} \text{ grains} \qquad (8.12) \\ &\sim 9 * 10^{30} \text{ grains}.\end{aligned}$$

How accurate is this answer? To see, think about where our estimates might be wrong. The radius of the earth is indisputable, and correct to two significant digits as given. The only question is in estimate 8.9. Perhaps 20 grains in a 1-cm row is too large. If 10 grains were the right number then estimate 8.9 would have to be replaced by

$$1 * 10^3 \text{ grains/cm}^3. \tag{8.13}$$

Then the answer in Equation 8.12 would be about 8 times too large; we would have to replace it by

$$1 * 10^{30} \text{ grains}. \tag{8.14}$$

If the correct estimate for grain size were 40 per cm, and there were

$$40^3 = 64 * 10^3 \text{ grains/cm}^3,$$

then the answer in Equation 8.12 would be 8 times too small and would have to be changed to

$$72 * 10^{30} \sim 7 * 10^{31} \text{ grains}. \tag{8.15}$$

The important observation here is that the answer in Equation 8.12 is the right *order of magnitude*. The number of grains of sand that would fill the earth is a number with 31 or 32 digits to the left of the decimal point.

EXAMPLE 5. ELECTRIC POWER CONSUMPTION

How much electric energy was used for domestic purposes (not including business or industry) in the United States in 1980?

Before we can answer this question we must think about the units in which electric energy is measured. A standard unit is the *watt-hour* (symbol Wh). It is the amount of energy that will keep a one-watt bulb lit for one hour. A more convenient unit is the kilowatt-hour (symbol kWh), which is 1000 watt-hours. One kilowatt-hour of energy will light a 100-watt bulb for 10 hours, or run a 1000-watt heater for 1 hour. A typical household might use about 500 kWh per month. Check your latest electricity bill to see how your own usage compares. It probably differs, but not by more than one order of magnitude. That is, it is probably more than 50 kWh but less than 5000 kWh. And so we can feel comfortable using $5 * 10^2$ kWh per household per month in our estimate. All we need now is an estimate of the number of households

Table 8.1 More metric prefixes

Prefix	Symbol	Meaning: multiply by
kilo	k	10^3
mega	M	10^6
giga	G	10^9
tera	T	10^{12}
milli	m	10^{-3}
micro	μ (Greek "mu")	10^{-6}
nano	n	10^{-9}
pico	p	10^{-12}
femto	f	10^{-15}
atto	a	10^{-18}

in the United States. The 1980 census found the population to be about 230 million, or $2.3 * 10^8$ people. If we figure that there are 4 people per household, then our final estimate is

$$5 * 10^2 \frac{\frac{\text{kWh}}{\cancel{\text{month}}}}{\cancel{\text{household}}} * 12 \frac{\cancel{\text{months}}}{\text{yr}} * \frac{1}{4} \frac{\cancel{\text{household}}}{\cancel{\text{person}}} * 2.3 * 10^8 \cancel{\text{people}}$$
$$= 3.5 * 10^{11} \text{ kWh/yr.}$$

The metric system has prefixes well beyond *kilo*, which means "times 10^3," to deal with such large numbers. Table 8.1 lists the prefixes for larger and smaller units. Using this information we can express our answer more compactly in several ways:

$$\begin{aligned} 3.5 * 10^{11} \text{ kWh} &= 3.5 * 10^8 \text{ MWh} \quad \text{(megawatt-hours)} \\ &= 3.5 * 10^5 \text{ GWh} \quad \text{(gigawatt-hours)} \\ &= 3.5 * 10^2 \text{ TWh} \quad \text{(terawatt-hours)} \end{aligned}$$

Thus our final estimate for domestic electric power consumption in the United States in 1980 is 350 terawatt-hours. The correct figure might be 100 TWh, or even 900 TWh, but is almost certainly not 35 TWh or 3500 TWh. We have the right number of zeros, that is, the correct ***order of magnitude***.

The second part of Table 8.1 lists the prefixes for expressing very small numbers. We discovered in Example 2 that it takes light $3.3 * 10^{-11}$ seconds to travel one centimeter. This number is more easily expressed as:

$$3.3 * 10^{-11}\text{ s} = 3.3 * 10^{-8}\text{ ms (milliseconds)}$$
$$= 3.3 * 10^{-5}\ \mu\text{s (microseconds)}$$
$$= 3.3 * 10^{-2}\text{ ns (nanoseconds)}$$
$$= 3.3 * 10^{1}\text{ ps (picoseconds).}$$

Light travels one centimeter in 33 picoseconds.

PROBLEMS

1. Compute these and write your answers in scientific notation:

*(a) $\dfrac{7.18 * 10^{15}}{5.33 * 10^{13}}$ (b) 0.0057

*(c) $\dfrac{9.37 * 10^{15}}{9.78 * 10^{-7}}$ (d) 680 000

*(e) $\dfrac{10^{-123} * 10^{11}}{10^{33}}$ (f) 1234.560

*(g) $(6.32 * 10^{61})^3$ (h) $\dfrac{1}{5 * 10^6}$

*(i) $0.079 * 10^5$ (j) $\dfrac{1}{5} * 10^6$

*(k) 0.000 400 (l) 0.004 00

*2. Estimate each of the following *without* a calculator. Then use your calculator to check. Write your answer in scientific notation if it is very large or very small.

(a) $\dfrac{4\,083\,692 * 49\,312}{2123 * 1945}$

(b) 0.000 342 9 * 0.000 000 966

(c) 0.000 000 79 * 3 240 726

*3. (a) Convert 0.000 036 8 mm to nm

(b) Convert $2.1 * 10^8$ fs to μs

(c) Convert 10^7 kW to GW

4. Write each of the following in scientific notation:

(a) $621 * 10^{111}$ (b) $621 * 10^{-111}$

(c) $\dfrac{1}{621} * 10^{111}$ (d) $\dfrac{1}{621 * 10^{111}}$

(e) $\dfrac{1}{621} * 10^{-111}$ (f) $\dfrac{1}{621 * 10^{-111}}$

(g) $-621 * 10^{111}$ (h) $-621 * 10^{-111}$

5. Practice the rules for exponents by simplifying each of the following where possible:

(a) $a^6 * a^7$	(b) $\frac{b^6}{b^7}$	(c) $c^6 + c^7$ (Careful!)
(d) $(d^6)^7$	(e) $e^6 * f^6$	(f) $f^6 * g^7$ (Careful!)
(g) $g^6 * g^{-6}$	(h) $\frac{h^6}{h^{-6}}$	(i) $i^a + i^b$
(j) $j^a + j^b$	(k) $(k^a)^b$	(l) $(m^a)^a$
(m) $(m^a)^{-a}$	(n) $(n^a)^{1/a}$	

6. A *byte* is roughly the amount of information required to specify one character on a typewriter keyboard:

$$a \text{ or } A \text{ or } b \text{ or} \ldots \text{or ! or} \setminus \text{or \#}\ldots$$

   (a) How many bytes of information are required to set this book in type? How many kilobytes? Megabytes?

   (b) Magnetic tape typically stores information at the rate of 800 bytes per inch, with a gap of 3/4 inch every 2^{10} bytes. How many inches of tape does it take to store the text of this book?

*7. A PDP–11/34 computer executes 0.75 million instructions per second.

   (a) How long does it take to execute one instruction? Answer in ms, μs, ns, or ps, whichever is most appropriate.

   (b) How many instructions can the computer execute in a year?

8. It is said to be impossible to solve the popular puzzle called the *magic cube* or *Rubik's cube* by trial and error because there are 43 252 003 274 489 856 000 possible positions.

   (a) Express this number in scientific notation, rounded to two significant digits.

   (b) How many years would it take to try out every position, at the rate of one per second?

   (c) Compare your answer in (b) to the age of the universe, which is estimated to be 15 billion years.

   (d) In the January, 1982 issue of *Scientific American* the state of the universe in the year 10^{119} is discussed. Would the Rubik's cube be solved by then?

   (e) Find and read Arthur C. Clarke's short story "The Nine Billion Names of God."

*9. When we have 9-digit (rather than 5-digit) zip codes, how many will each of us be able to have all to ourselves?

10. Christopher Marlowe wrote that Helen's was "the face that launch'd a thousand ships." The beauty of everyday faces is better measured in millihelens. How many ships could a 3-millihelen face launch?

11. The United States gross national product for 1980 was about $2400 billion.

(a) Express this number in scientific notation.

(b) Estimate the height of a stack of $2400 billion worth of $100 bills. Express your answer in cm, m, or km, whichever is appropriate. (Start by measuring a short stack of bills. Dollar bills will do.)

(c) Answer question (b) for a stack of one-dollar bills and for a stack of pennies.

12. Find out the current United States national debt. How much of that debt is each of us responsible for?

*13. A new kind of electron microscope has recently been used to study microtubules, which are only 6 nanometers thick. The thinnest strand that can be seen by the human eye is about 0.05 mm. By what factor must the microtubules be magnified in order to make them just visible?

14. The universe is thought to be a sphere with radius $4 * 10^9$ (4 billion) light-years. (The question of what lies beyond that sphere belongs to philosophy.) Use the formula for the volume of a sphere in terms of its radius to calculate the volume of the universe, in cubic light-years; in cubic kilometers; and in cubic centimeters.

*15. One gram of hydrogen consists of $6 * 10^{23}$ atoms. What does one atom of hydrogen weigh?

16. Matter in the universe is mostly hydrogen. It is pretty sparsely distributed. On the average there are only

$$4 * 10^{-31} \text{ g/cm}^3$$

(a) Use the data from the previous problem to find out how many atoms of hydrogen there are in a cubic centimeter. Your answer should be much less than 1. Write it as a decimal as well as in scientific notation.

(b) How many cubic centimeters of space do you think you would have to examine before you found one that wasn't empty?

(c) What is the density of matter, in atoms per cubic kilometer?

(d) Use the answers to part (a) and to Problem 14 to estimate the number of atoms in the universe.

17. The number 10^{100} is known as a *googol.*

(a) Is the largest number your calculator can handle larger or smaller than a googol?

(b) Are there a googol of atoms in the universe? (See the preceding problem.)

(c) A *googolplex* is 10^{googol}, that is, 1 followed by a googol of zeros. Estimate how much space you would need to write out this number. You should plan on making the zeros very small. Even then the answer may surprise you. Part (b) may explain why.

18. Estimate the number of gallons of gasoline burned by private automobiles in the United States in a year. Compare your answer with that resulting from Problem 11 of Chapter 5.

19. Estimate the number of bricks used to construct some large brick building near you.

20. Estimate the number of quarts of milk consumed in the United States in a year.

21. Estimate the number of blades of grass on a football field.

22. How many times do your eyes blink in a year?

23. How many times does a heart beat in a lifetime?

24. Figure 8.1 is a sketch map of Eastern Massachusetts. Six feet of snow fell in 24 hours in the Great Blizzard of 1978. Estimate the number of snowflakes that fell on the roughly semicircular area bounded by Route 128 and the coastline. First estimate the area covered, in square meters. Then use one of two methods: the first starts with an estimate of the number of snowflakes/cm^3, the other with an estimate of the number of snowflakes that fall per minute or per second on a square meter.

25. Estimate the number of leaves that fall from the trees annually in your state. Use either of the two methods of the previous problem.

26. Estimate the number of secondary schools in the United States.

27. Estimate the number of ice cubes that could be made from the ice cap at the North Pole.

*28. Einstein's theory of relativity correctly predicted that energy is

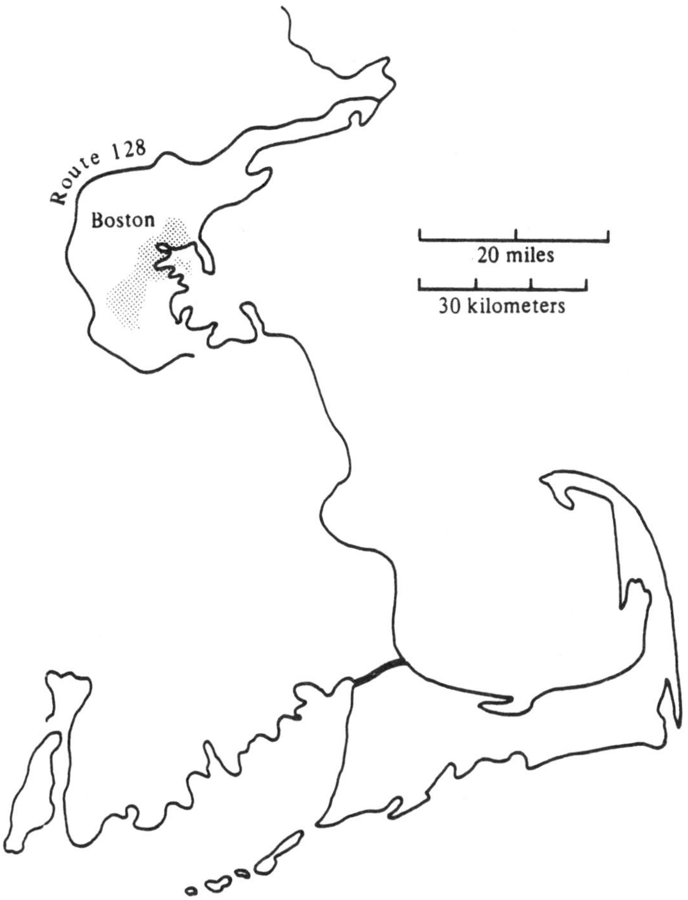

Figure 8.1 Sketch map of Eastern Massachusetts

a special form of mass. The conversion factor is enormous. It is the *square* of the speed of light:

$$9 * 10^{20} \text{ cm}^2/\text{sec}^2$$

The speed of light is traditionally denoted by the letter c, hence the famous formula:

$$E = mc^2,$$

which expresses the direct proportionality between mass m, measured in grams, and energy E, measured in dyne-centimeters. The dyne-centimeter is not a common, everyday unit of energy.

Here is how you can convert it into more familiar terms. First, dyne-centimeters can be coverted to watt-seconds:

$$10^7 \text{ dyne-centimeters} = 1 \text{ watt-second.}$$

Then watt-seconds can be converted to kilowatt-hours.

(a) How many watt-seconds per kilowatt-hour?

(b) How many dyne-centimeters per kilowatt-hour?

(c) If one gram of matter were completely converted to energy how many kilowatt-hours of energy would be produced?

(d) How long would that much energy keep a 100-watt lightbulb burning? Is your answer best expressed in hours, days, years, or millennia?

29. According to the *Boston Globe* (12/27/80), an unpublished Department of Energy study said that nuclear power has benefited since its beginning from federal aid in five major areas: $23.6 billion for research and development, $237.4 million to promote foreign reactor sales, $2.5 billion for uranium market promotion, $7.1 billion in fuel enrichment pricing aid, and $6.5 billion for management of wastes, mining spoils cleanup, and unpaid decommissioning costs.

(a) Which of these numbers is negligible compared to the others?

(b) Find the total federal subsidy to the nuclear power industry through 1980.

*30. The eruption of Mount St. Helens on May 18, 1980 released about $1.7 * 10^{18}$ joules of energy. (A *joule* is another name for a watt-second. We studied watt-seconds in Problem 28.) The detonation of a one-megaton nuclear bomb releases about $4 * 10^{15}$ joules.

(a) How many one-megaton bombs would have to be exploded to release the same amount of energy as did the eruption of Mount St. Helens?

(b) Are there that many bombs in the United States arsenal?

(c) What does the phrase *one-megaton* in the expression *one-megaton bomb* mean? The atomic bombs dropped in World War II were kiloton bombs. How many kiloton bombs would it take to make a one-megaton explosion?

CHAPTER 9

Exponential Growth

I remember as a boy discovering the delights of multiplication and calculating that about 600 years ago I had more ancestors than there were people in the world. I must confess this still unnerves me a little, just like the reflection that a penny put out to accumulate interest at any reasonable rate will equal all the capital in the world in a few hundred years.

– Kenneth Boulding

When to multiply rather than add / Exponential growth and decay / Markups and markdowns / More review of rules for manipulating exponents / Exponentiation with the calculator

EXAMPLE 1. INFLATION

A television set that sold for \$94.50 a year ago now retails for \$106.50. If inflation continues at the same rate, what will happen to the price of the new television as the years go on?

Our first step, as usual, is to identify the variables. The price P in dollars of a new television set depends on Y, the year of purchase. We will start counting years when we first collected data: namely, last year. Thus P = \$94.50 when $Y = 0$, and now $Y = 1$ and $P(1) = \$106.50$. Next year will correspond to $Y = 2$.

So far we have studied only linear dependence. If P depended linearly on Y we would know what to do. The line's slope would be

$$\frac{106.50 - 94.50}{1 - 0} = 12 \text{ \$/yr.} \tag{9.1}$$

Its intercept would be \$94.50, and so the equation would be

$$P = 94.50 + 12Y. \tag{9.2}$$

The graph of this equation is the dashed line in Figure 9.1. But this mathematical model for inflation is not realistic. The truth is worse: it is the curve in Figure 9.1. When you read about increasing prices, the units of the increase are percentages per year rather than dollars per year. The proper question is: "What do we *multiply* last year's price by to get this year's?" rather than: "What do we *add* to last year's price to get this year's?" The proper answer is the ratio (new price)/(old price):

$$\frac{P(1)}{P(0)} = \frac{P(\text{when } Y = 1)}{P(\text{when } Y = 0)} = \frac{106.50}{94.50} = 1.127. \tag{9.3}$$

The number 1.127 is called the *annual growth factor*. It is what you *multiply by* each year to find the new price. The growth factor is

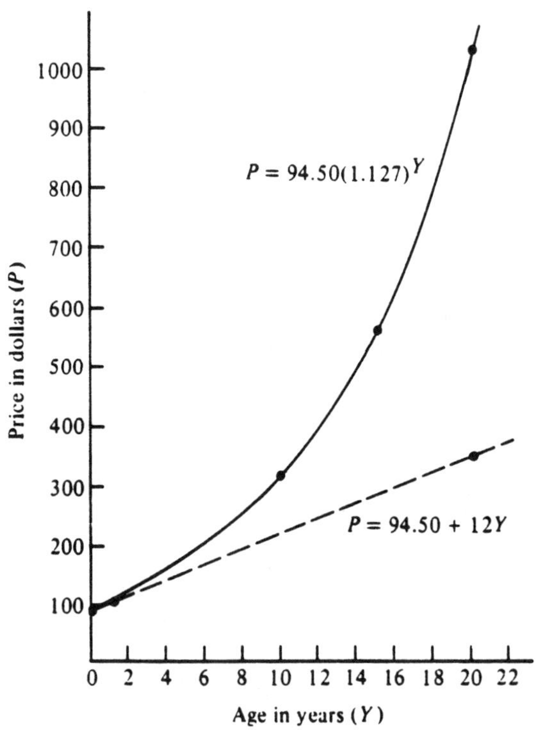

Figure 9.1 Inflation

analogous to the slope in the linear model; the slope is what you *add* each year. Use the growth factor 1.127 to find next year's price:

$$P(\text{next year}) = P(\text{now}) * 1.127 \qquad (9.4)$$

so

$$P(2) = 106.50 * 1.127 = \$120.02.$$

Notice that this price is larger than that predicted by the linear model, which is only

$$P(2) = 94.50 + 12 * 2 = \$118.50.$$

Let us look at Equation 9.5 a little more closely. It says that

$$\begin{aligned} P(2) &= P(1) * 1.127 \\ &= P(0) * 1.127 * 1.127 \\ &= P(0) * 1.127^2 \\ &= 94.50 * 1.127^2. \end{aligned} \qquad (9.5)$$

The argument in Equation 9.6 keeps right on going:

$$P(3) = 94.50 * 1.127^3 = 135.26 \qquad (9.6)$$

and, in general,

$$P(Y) = 94.50 * 1.127^Y. \qquad (9.7).$$

Table 9.1 gives the results of calculating P using Equation 9.7.

Table 9.1 How prices increase at 12.7% inflation

Y (years)	P (Price of new TV set, \$)
0	94.50
1	106.50
2	120.02
3	135.26
5	171.80
10	312.33
15	567.80
20	1032.26

> To check Table 9.1 you will have to learn to use the exponentiation, or [y-to-the-power-x] key on your calculator. Start by verifying the result in Equation 9.6. To cube 1.127, try a sequence like
>
> $$1.127\ [y^x]\ 3\ [=].$$
>
> (Details depend on the calculator you own.) Then learn where the parentheses go if you need to use them to finish checking Equation 9.6. That equation tells you to "cube 1.127, then multiply by 94.50"–not to "multiply 94.50 by 1.127 and cube the answer." If your answer differs a little from the one in the text it may be because you entered the rounded annual growth factor 1.127 instead of calculating and storing the true value of the ratio 106.50/94.50.

The results listed in Table 9.1 are graphed in Figure 9.1. The curve there, and Table 9.1, and Equation 9.7, all express the same *exponential model* for inflation. Compare it with the linear model on the graph, which represents Equation 9.1. Each curve connects the points (0, 94.50) and (1, 106.50). The line starts out above but is soon well below the curve. To follow the line we *add* a fixed amount annually–an amount equal to the slope. To follow the curve we *multiply* each year by the annual growth factor. The growth factor is not an amount, it is a proportion. Its units are dollars per dollar ($/$).

EXAMPLE 2. POPULATION GROWTH

A colony of bacteria has an hourly population growth factor of 1.7. Let us see how it grows.

Suppose we measure the size of the colony by finding its weight W in milligrams. W grows exponentially: each hour we *multiply* its value by 1.7 to find out what happens. If the weight at noon is 3.8 mg then at 1 P.M. it will be

$$3.8 * 1.7 = 6.46 \sim 6.5 \text{ mg}.$$

The weight is given by

$$W(T) = 3.8 * 1.7^T, \tag{9.8}$$

where T is the number of hours since noon.

To find out what happens in a day, substitute $T = 24$ in Equation 9.8:

$$\begin{aligned} W(24) &= 3.8 * 1.7^{24} \\ &= 3.8 * 339\,000 \\ &= 1\,290\,000 \qquad (9.9) \\ &\sim 1.3 * 10^6 \text{ mg} \\ &\sim 1.3 * 10^3 \text{ kg.} \end{aligned}$$

Our 3.8-mg colony has grown to more than a Megagram in a day. The figure 339 000 above is the *daily* growth factor, which is not just 1.7 * 24. Since each hour W is multiplied by 1.7, in a day it will be multiplied by 1.7 twenty-four times, that is, by $1.7^{24} = 339\,000$. The magnitude of that growth factor accurately portrays the speed of unchecked exponential growth.

The growth factor for T hours is just

$$\text{g.f. for } T \text{ hours} = (\text{hourly g.f.})^T = 1.7^T. \qquad (9.10)$$

To find the 2-hour growth factor, substitute $T = 2$ in Equation 9.10 to get 2.89. In two hours the colony nearly triples in size.

How much did the colony weigh at 11 A.M., one hour before it weighed 3.8 mg? We shall figure this out using first common sense and then Equation 9.10, and see that the answers agree. Common sense tells us that, because we *multiply* by 1.7 for each hour that passes, to know what happened an hour ago we should *divide* by 1.7. Thus at 11 A.M. the colony weighed

$$\frac{3.8}{1.7} = 2.24 \text{ mg.}$$

To use Equation 9.10 to answer the same question, we observe that when T is measured starting at noon, 11 A.M. corresponds to $T = -1$ hour. Substituting in Equation 9.10 gives

$$W(-1) = 3.8 * 1.7^{-1} = 2.24 \text{ mg.}$$

> To check this on your calculator, you should use the change sign key, [CHS] or [+/−]. A sequence like
>
> 1.7 [y^x] 1 [+/−] [=]
>
> calculates 1.7^{-1}. Then multiply the result by 3.8 mg.

The two answers agree because

$$1.7^{-1} = \frac{1}{1.7}.$$

This illustrates the rule for negative exponents:

$$K^{-E} = \frac{1}{K^E}. \tag{9.11}$$

We encountered this rule in Chapter 8 for powers of $K = 10$. This rule actually makes things easier, not more difficult. It allows us to use Equation 9.8 when T is negative, to study the past as well as the future.

Equation 9.8 even gives the right answer in the present. At noon, $T = 0$. Then

$$W(0) = 3.8 * 1.7^0 = 3.8 \text{ mg},$$

because

$$1.7^0 = 1.$$

The rule that

$$K^0 = 1, \tag{9.12}$$

which may seem odd at first, is the only possible rule that allows Equation 9.8 to give the correct result when $T = 0$.

What is the weight of the culture at 12:30? To find out, we must multiply the weight at noon, 3.8 mg, by the half-hour growth factor. Equation 9.10 says that the T-hour growth factor is 1.7^T. When $T = 1/2 = 0.5$ the calculator gives

$$W(0.5) = 1.7^{0.5} = 1.3038 \sim 1.3$$

for the half-hour growth factor. The weight at 12:30 must then be

$$3.8 * 1.7^{0.5} = 4.954 \sim 5.0 \text{ mg}.$$

It makes sense that the half-hour growth factor of 1.3 is between 1 and 1.7. But notice that it is *not* 1.35, which would be halfway between.

Be sure you understand also that the half-hour growth factor is *not* half of the one-hour growth factor. That would be 0.85. To understand the half-hour growth factor, think about what happens in each of the two half-hour periods between 12 and 1 o'clock. The half-hour growth factor gets two chances to work, so

$$(\text{half-hour g.f.})^2 = \text{1-hour g.f.} = 1.7. \tag{9.13}$$

(Check that $1.3038^2 = 1.7$.) Equation 9.13 shows that the half-hour growth factor is the *square root* of the hourly growth factor 1.7. (Use your calculator to check that $\sqrt{1.7} = 1.3038$.) The seemingly strange rule that

$$K^{1/2} = \sqrt{K} \tag{9.14}$$

is forced on us if we are to have Equations 9.8 and 9.10 give correct answers when $T = 1/2$.

EXAMPLE 3. RADIOACTIVE DECAY

Exponential curves, like lines, can fall as well as rise. The radioactive isotope iodine-131 happens to be dangerous. Fortunately it doesn't last long. If you start with 1.235 grams, a day later you will have only 1.133 grams left. Let us see how the quantity of iodine-131 decays.

To find the daily growth factor, divide:

$$\frac{\text{amount after 1 day}}{\text{original amount}} = \frac{1.133}{1.235} = 0.9174. \tag{9.15}$$

In this case the growth factor might better be called a shrinkage factor, since each day you have less than the day before. A daily growth factor that, like 0.9174, is *less than 1*, corresponds to shrinkage just as, for straight lines, a negative slope corresponds to a falling line. The formula relating the amount W of iodine-131 in grams to the number of days D is

$$W = 1.235 * 0.9174^D. \tag{9.16}$$

Table 9.2 gives some values of W and Figure 9.2 shows the graph of Equation 9.16. You should check both.

Table 9.2 Decay of iodine-131

D (days)	W (grams)
0	1.235
1	1.133
2	1.039
5	0.803
10	0.522
20	0.220
30	0.093

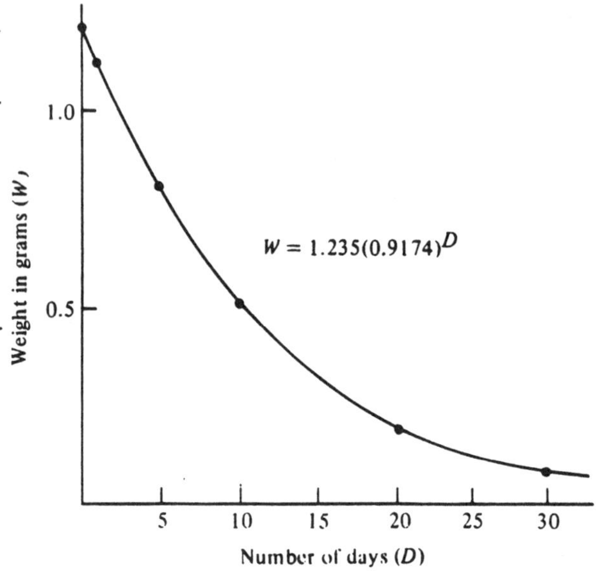

Figure 9.2 Decay of iodine-131

EXAMPLE 4. A COMPARISON OF LINEAR AND EXPONENTIAL GROWTH

If a quantity Q depends linearly on another quantity T, we know that the equation relating Q to T has the form

$$Q = \text{intercept} + \text{slope} * T. \tag{9.17}$$

The units of the intercept are those of Q; the units of the slope are (units of Q)/(units of T). But suppose the dependence is exponential. We found in the preceding examples that the equation relating Q to T now takes the form

$$Q = \text{intercept} * (\text{growth factor})^T. \tag{9.18}$$

The units of the intercept are those of Q; the units of the growth factor are (units of Q)/(units of Q). Thus the growth factor has no units; it is a simple proportion.

We can use Equations 9.17 and 9.18 to construct Table 9.3, which compares linear and exponential models. The first three lines come directly from Equations 9.17 and 9.18. To convert the former equation to the latter, we replace addition (+) by multiplication (*), multiplication (*) by exponentiation, and slope by growth factor. (There is no

Table 9.3 Linear and exponential models compared

	Linear	Exponential
1.	+	*
2.	*	exponentiate
3.	slope	growth factor
4.	0	1
5.	positive	greater than 1
6.	negative	less than 1
7.	−	÷ or /
8.	half of	square root of

standard symbol for exponentiation. We indicate this operation by writing raised exponents. Computer languages use either "**" or " ".) If you understand the first line, then the second makes sense: multiplication (*) is repeated addition (+), exponentiation (indicated by superscripts) is repeated multiplication (*). The first two lines of the table can also help you remember the first rules for exponents. In

$$K^{B+C} = K^B * K^C,$$

you replace + by *, because adding the exponents B and C corresponds to multiplying the numbers K^B and K^C. In

$$K^{B*C} = (K^B)^C,$$

you replace * by exponentiation, because multiplying B and C in the exponent corresponds to raising K^B to the power C.

The analogy in the third line of the table is consistent with the first line. In linear growth you *add to Q* the value of the slope each time period (year, day, hour, or second); in exponential growth you *multiply Q by* the growth factor. After T periods of time have gone by, you either add (slope) * T (in a linear model) or multiply by (growth factor)^T (in an exponential model). The fourth, fifth, and sixth lines of Table 9.3 make sense if you think of them as describing the slope or growth factor. In the linear model, when the slope is 0, Q does not change, because $Q + 0 = Q$. In the exponential model, when the growth factor is 1, Q does not change, because $Q * 1 = Q$. The fourth line will help you remember the rule

$$K^0 = 1.$$

When the slope is positive, Q is increasing. When the growth factor is greater than 1, Q is increasing (see Examples 1 and 2). If Q depends linearly on T, Q decreases as T grows when the slope is negative. If Q depends exponentially on T, Q decreases as T grows when the growth factor is less than 1 (see Example 3). The seventh line of the table is a natural consequence of the first. Subtraction undoes addition and division undoes multiplication. This line is linked to the rule

$$K^{-B} = \frac{1}{K^B},$$

in which subtraction (—) is converted to division (/).

The last line of the table sometimes causes confusion. Think of half of K as the solution to the equation

$$X + X = K.$$

Here, X is what you *add* to itself to get K. The corresponding equation on the exponential side of the table is

$$X * X = K.$$

Here, X is what you *multiply* by itself to get K. The last line of the table will help you remember that

$$K^{1/2} = X.$$

EXAMPLE 5. GROWTH FACTORS AND PERCENTAGES

In successive years tuition at ESU increased by 12 percent, then 15 percent. What was the overall percentage of increase?

Please do not jump to the conclusion that the answer is 12 + 15 = 27%. Percentages are proportions, not amounts. You can add amounts, but, as we shall see, you must multiply proportions. (That is why we include this example in this chapter.) Suppose the tuition before the first increase was $2000. Then the increase was 12% of 2000, or 2000 * 0.12, and so the new tuition, after the first increase, was

$$\begin{aligned} \text{New tuition} &= \text{old tuition} + \text{increase} \\ &= 2000 + 2000 * 0.12 \\ &= 2000\,(1 + 0.12) \qquad (9.19) \\ &= 2000(1.12) \\ &= \$2240. \end{aligned}$$

The transition from the second to the third lines above is a little subtle. Where does the 1 come from? It is there all the time, but not expressed in the second line, where 2000 is the same as 2000 * 1. We choose to make it visible in order to arrive at the fourth line, which looks familiar. It tells us that

a 12% increase corresponds to a growth factor of 1.12.

This is a special case of the following general principle, which we will explore in this example:

$$\text{An } r\% \text{ increase means a growth factor of } 1 + \frac{r}{100}. \qquad (9.20)$$

And so, for example, the 15% increase at ESU corresponds to a growth factor of 1.15. To find the tuition after the second increase, we calculate

$$\begin{aligned} \text{tuition after second increase} &= \text{tuition after first increase} * 1.15 \\ &= 2000 * 1.12 * 1.15 \\ &= 2000 * 1.288. \end{aligned}$$

The combined growth factor from the two increases is thus 1.288, which corresponds to an increase of 28.8%, or nearly 29%. The difference between that figure and the sum, 27%, is 1.8%. That 1.8% appears because the 15% increase applies to the 12% increase as well as to the original $2000: 1.8% is 15% of 12%.

If we look back to Example 1 we can now see that the annual growth factor of 1.127 corresponds to an inflation rate of 12.7% per year. In Example 2 the hourly growth factor of 1.7 means a 70% increase in bacteria each hour.

When tuition goes up 100%, it doubles. The principle in 9.20 still applies. When $r = 100\%$, the growth factor is

$$1 + \frac{100}{100} = 2.$$

A growth factor of 2 is indeed doubling.

The same principle works for percentage decreases, too. If a manufacturer offers a 13.5% rebate on a $450 refrigerator, the true cost is

$$\begin{aligned} 450 - 13.5\% \text{ of } 450 &= 450 - 450 * 0.135 \\ &= 450 * (1 - 0.135) \\ &= 450 * 0.865. \end{aligned}$$

The growth factor is

$$1 + \frac{-13.5}{100} = 0.865.$$

The negative sign is there because the true cost is 13.5% less than the original cost. It is thus 86.5% of this cost. Notice that the fact that the refrigerator originally cost $450 has not entered the calculation. The iodine-131 in Example 3 shrinks at the rate of 8.26% per day.

Finally, we can use principle 9.20 to understand how people in business compute profits. If Wing Aero manufactures solar panels for $375 and wishes to make a 23% profit based on that cost, the selling price will have to be 375 * 1.23 = 461.25. The dollar profit is the difference, $86.25, which is indeed 23% of 375. If they wish to appear less greedy, they can report the profit as a percentage of the selling price: $86.25 is only 18.7% of $461.25. The relation between the 23% profit (based on cost) and the 18.7% profit (based on selling price) can be calculated directly from principle 9.20. The growth factor for calculating the selling price based on the cost is 1.23. Hence the growth factor for calculating the cost based on the selling price is 1/1.23 = 0.813. The difference, 1 − 0.813, is just the 18.7% profit based on the selling price. Notice again that the dollar cost of the panels does not enter the calculation at all.

PROBLEMS

1. Compute:

$$1.127^{20} \tag{9.21}$$

with your calculator in two ways: first as

$$1.127 * 1.127 * 1.127 * \ldots \text{ (20 times).}$$

(Use the memory register to store 1.127.) Second, use the exponentiation key [y^x] to compute 1.127^{20} directly. You should get the same answer. Is this the same as 1.127 twenty times? That is, is it the same as 1.127 times 20? Calculate 1.127 * 20 to see.

2. Check to see that your calculator knows the rules and that you know how to use your calculator by verifying the following equations. Store 1.82 in the memory to save time.

(a) $1.82^0 = 1$ (b) $1.82^{0.0001} = 1.000\,06$

(c) $1.82^1 = 1.82$ (d) $1.82^{1.0001} = 1.82 * 1.82^{0.0001}$

(e) $1.82^{-1} = \frac{1}{1.82}$ (f) $1.82^{0.5} = \sqrt{1.82}$

(g) $1.82^{-0.5} = \frac{1}{\sqrt{1.82}}$ (h) $1.82^{-4.5} = \left(\frac{1}{1.82}\right)^{4.5}$

(i) $(1.82^2)^3 = 1.82^6$ (j) $(1.82^6)^{(1/6)} = 1.82$

(k) $1.82^{5.9} * 1.82^{1.7} = 1.82^{(5.9+1.7)}$

(l) $1.82^{8.3} * (1.82^4)^{-1.3} = 1.82^{(8.3-4*1.3)}$

*3. Simplify each of the following, when possible:

(a) $2^A 2^B$ (b) $2^A 2^{-B}$ (c) $3^{2A} 3^{-A}$
(d) $3^{2A} 3^{1/A}$ (e) $5^{(A^2)} 5^{-A}$ (f) $5^{(A^2)} 5^{1/A}$
(g) $(7^A)^2$ (h) $(7^A)^A$ (i) $8^{2/3}$
(j) $16^{3/2}$ (k) $11^E 13^E 17^E$ (l) $19^{2E} 23^E$
(m) $(29^{3F})^2$ (n) 29^{6F} (o) $(31^I)^J - (31^J)^I$
(p) $\frac{(31^I)^J}{31^{IJ}}$ (q) $s^t s^{-t}$ (r) $\frac{t^u}{t^{-u}}$

*4. (a) Estimate from Figure 9.1 the time Y at which the price of the television set will be twice what it is in year 0, according to each model: linear and exponential.

(b) Use algebra to find the doubling time exactly for the linear model.

(c) Try to find the doubling time numerically for the exponential model by trial and error, using your calculator.

5. Prepare a table and draw a graph illustrating the growth of the bacteria colony in Example 2. Your data should cover a twenty-four-hour period.

6. Sales of the Blue Gear Co. were $2.11 million in 1978 and $2.34 million in 1979.

(a) Suppose first that sales grow linearly with time in years since 1978. Name the variables, find the equation relating them, and graph it. Predict sales for 1980, 1985, and 2000. When will annual sales reach $3 million?

(b) Now, using the same data for 1978 and 1979, suppose sales grow exponentially. Find the annual growth factor. Find the equation (like Equation 9.8) that tells how sales depend on time. Use it to prepare a table predicting sales in 1980, 1981, 1983, 1985, 1990, and 2000. Graph the exponential growth curve on the same set of axes as the line showing linear growth. When does the exponential model predict that sales will reach $3 million?

*7. Calculator prices have been falling in spite of inflation. One that sold for \$85.50 a year ago sells for \$73.75 now. Suppose the drop in prices is exponential.

(a) Find the annual growth factor (which should be less than 1). By what percentage does the price fall in a year?

(b) Name the variables and write an equation (like Equation 9.16) in the text. Prepare a table and use it to draw a graph predicting what will happen to the price of the calculator for the next twenty years. Estimate when the calculator will sell for \$50.

8. Suppose a week's worth of groceries cost \$75 in 1980. Write formulas giving the cost of these groceries Y years from now if inflation is 9%/year (annual growth factor = 1.09), and then if it is 13%/year. Draw the exponential growth curves on the same set of axes. Predict in each case when the cost of the groceries will have doubled.

*9. If the inflation rate is 1.3% per month what is it per year?

10. The Dutch bought Manhattan Island in 1626 for \$24 worth of merchandise.

(a) If the same amount of money had been invested in an account paying 8% interest each year (since each year's interest is reinvested, the annual growth factor is 1.08) what would it be worth now?

(b) Estimate the present worth of Manhattan and decide whether the Dutch made the right move.

*11. The population of the world was about 4 billion in 1980. For some time it has been growing at about 2% per year. Find an exponential formula for the population. How many people are there now? How many were there in 1900? How many will there be in the year 2000? In the year 3000?

12. Twenty-two rabbits were brought to Australia in 1860. A year later there were 95.

(a) Find a formula showing how the number of rabbits R depends on the time in years Y since 1860 under each of the following assumptions:

(L) The rabbit population grows linearly.

(E) The rabbit population grows exponentially.

The second assumption is the more reasonable since rabbits don't add, they multiply!

(b) Graph the two growth curves on the same set of coordinate axes. (The exponential graph is hard to fit on the page when you use a proper vertical scale!)

(c) For each model, answer the following questions: What was R in 1861? In 1862? In 1863? In 1890? In 1920? In 1940? When did the rabbit population reach 1000? 1 000 000? (This problem anticipates Example 5 of Chapter 11.)

*13. In 1980 about 170 million cars and trucks were registered in the United States. The number is growing at a rate of 4% per year.

(a) How many were there in 1974, when the public first noticed an oil crisis? How many will there be in the year 2000?

(b) What is the number of cars and trucks *per person*?

(c) If the population grows at a rate of 2.2% annually, how does the number of vehicles per person change?

14. The metabolic rate M (measured in Calories/day of energy consumption) of a cold-blooded animal doubles for each 10°C increase in environmental temperature T. For a certain lizard, $M = 1200$ Cal/day at $T = 20°C$.

(a) Find M at $T = 30°$, $T = 40°$, and $T = 10°C$.

(b) Draw a graph showing how M depends on T.

(c) Check that the formula

$$M = 298.7(1.072)^T$$

gives the same values, to two decimal places, at the four temperatures. By examining the formula, determine the growth factor for 1°C. By what percentage does M change for each 1°C increase in environmental temperature?

15. (a) Explain why the number of ancestors you have has a growth factor of approximately 2 for each 25 years you look back into the past. What would the world's population have to have been 600 years ago in order for the observation that opens this chapter to be true?

(b) Check Boulding's second observation. Use the fact that 6% annual interest corresponds to an annual growth factor of 1.06.

*16. An executive's annual salary is $52 000. First she gets a 12% raise, then a 9.3% raise.

(a) What is her new salary?

(b) What single percentage raise would have had the same result?

(c) Explain why the answer to (b) is *not* 12% + 9.3% = 21.3%.

17. The boss says: "Hard times have come. Everyone takes a 10% pay cut." Then he says: "Prosperity returns. Everyone gets a 10% pay raise. Now we are all even, okay?" Is it okay? What if the raise preceded the cut?

*18. An appliance retailer sold sixty radios at $30.00 each after marking them down 20% of their selling price. All the radios were bought at the same price and had been marked up 50% on cost. What was the gross margin percentage earned on the original cost? On the final selling price?

19. A housewares manufacturer produces a can opener at a cost of $1.80. He sells it through a manufacturer's agent directly to large retailers. The agent receives a 20% commission on his sales, the retailers earn a margin of 30% on cost, and the manufacturer plans a net profit of 10% on his selling price. What is the retail price of the can opener?

*20. The attrition rate at ESU is 17% in the freshman year and 4% in each of the three succeeding years. What percentage of the incoming class graduates in 4 years? (Many of the students who withdraw subsequently return to finish their education. They are not considered in these statistics.)

CHAPTER 10

The Exponential Curve Through Two Points

How can it be that mathematics, being after all a product of human thought which is independent of experience, is so admirably appropriate to the objects of reality?

– Albert Einstein, *Geometry and Experience*

Finding the equation of the exponential curve joining two points / Fractional powers
Solving exponential equations by trial and error

EXAMPLE 1. CITYVILLE

Cityville was a boom town in the fifties. Its population was 116 000 in 1952 and 171 000 in 1957. Find the annual growth factor K and an exponential formula

$$P = \text{intercept} * K^T \tag{10.1}$$

for the population P. Measure T in years since 1950.

If the growth were linear we would first find the slope. Since it is exponential, we need the annual growth factor K instead. The data tell us that the

$$\text{5-year growth factor} = \frac{171\,000}{116\,000} = 1.47. \tag{10.2}$$

How are K and 1.47 related? We know from Equation 9.10, or directly from Equation 10.1, that K^T is the T-year growth factor. Thus K^5 is the 5-year growth factor and

$$K^5 = 1.47. \tag{10.3}$$

Equation 10.3 tells us that K is the fifth root of 1.47. All we need to do is find K. We do this by raising each side of Equation 10.3 to the 1/5 power and using the rule:

$$(K^B)^C = K^{B * C}.$$

Then

$$\begin{aligned} 1.47^{1/5} &= (K^5)^{1/5} \\ &= K^{(5 * 1/5)} \\ &= K^1 \\ &= K. \end{aligned}$$

Now find the value of $1.47^{1/5}$, using the [y^x] and [$1/x$] keys, and parentheses if necessary:

$$K = 1.08.$$

The population of Cityville is growing at a rate of 8% per year. In 5 years it grows 47% (*not* 40%).

To complete Equation 10.1 we must find the intercept: the value $P(0)$ which P has when $T = 0$, in 1950. We know that $K = 1.08$, and that $P(2) = 116\,000$. Substituting in Equation 10.1 gives

$$116\,000 = \text{intercept} * 1.08^2$$

so

$$\begin{aligned} P(0) = \text{intercept} = \frac{116\,000}{1.08^2} &= 116\,000 * 1.08^{-2} \\ &= 99\,000 \text{ people.} \end{aligned}$$

The exponential formula we seek is thus

$$P = 99\,000 * 1.08^T.$$

Had we saved all the arithmetic in our calculation of the growth factor we would have found the equation:

$$\text{growth factor} = \left(\frac{P(7)}{P(2)}\right)^{\left(\frac{1}{7-2}\right)} \tag{10.4}$$

This is analogous to finding the slope when the growth is linear using the equation:

$$\text{slope} = [P(7) - P(2)] * \frac{1}{7-2}. \tag{10.5}$$

To arrive at Equation 10.4 from Equation 10.5 we change one minus sign (—) to division (/) and a multiplication (*) to exponentiation, as suggested in Table 9.3.

> The computation of $1.47^{(1/5)}$ is the first we have encountered that we cannot do without the calculator. For the addition, subtraction, multiplication, and division the calculator has been a time saver, but not essential. Even large powers, like 1.127^{20}, or square roots, we could have managed. But calculations like $1.47^{(1/5)}$, or, worse, $1.47^{2.35}$, are hard to imagine doing by hand. Until very recently, people who had to make such calculations used a slide rule for approximate answers or, for more accuracy, logarithms, which we will study for other purposes in the next two chapters. The calculator has brought such computations within reach. But even a calculator has to work at them. The results of +, −, *, /, and $\sqrt{}$ seem to appear instantaneously on your machine, but exponentiation takes a noticeable length of time. A calculator approaches exponentiation by finding approximations to the right answer until it has one that is as accurate as the number of places it can display. That is, it calculates 3 * 3 directly, but 3^2 more slowly, by sneaking up on the answer, 9. If you study enough mathematics to reach the second semester of calculus you may then get an idea of how your calculator exponentiates.

EXAMPLE 2. DEPRECIATION

Wing Aero's second panel stamping machine was worth $32 000 new. It is now 6.5 years old, and it is worth $12 000. Compare exponential and linear models for its depreciation.

The variables in this problem are V, the value of the machine, which we'll measure in thousands of dollars, and Y, its age in years. $Y(0)$, the intercept, is 31 in both models. To calculate with the linear model we find the slope:

$$\text{slope} = \frac{V(6.5) - V(0)}{6.5} = \frac{12 - 31}{6.5} = -2.9\ \frac{\text{thousand \$}}{\text{year}}. \tag{10.6}$$

Thus the equation is

$$V = V(0) - \text{slope} * Y = 31 - 2.9Y. \tag{10.7}$$

The graph is shown in Figure 10.1.

The equation for the exponential model has the form

$$V = V(0) * K^Y = 31 * K^Y,$$

where K is the annual growth factor. Since V decreases with time, K

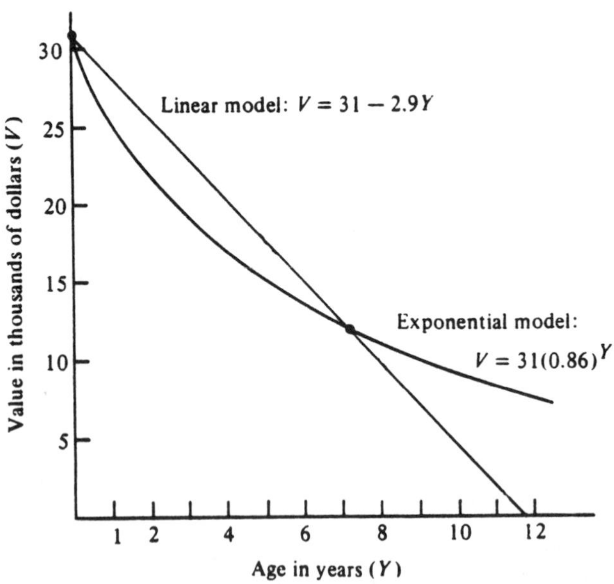

Figure 10.1 Two models for panel stamper depreciation

will be less than 1, just as the slope in Equation 10.7 is less than 0. To find K, we start with the 6.5-year growth factor, which is given by:

$$\frac{V(6.5)}{V(0)} = \frac{12}{31} = 0.3871.$$

Since the 6.5-year growth factor is also $K^{6.5}$, we know that

$$K^{6.5} = \left(\frac{V(6.5)}{V(0)}\right) = 0.387. \tag{10.8}$$

Next we must raise each side of Equation 10.8 to the power 1/6.5, which gives:

$$\begin{aligned} K &= \left(\frac{V(6.5)}{V(0)}\right)^{1/6.5} \\ &= (0.387)^{1/6.5} \\ &= 0.864147 \\ &\sim 0.86. \end{aligned} \tag{10.9}$$

The calculation in Equation 10.9 finds the growth factor just as the one in Equation 10.6 finds the slope. We now know that the equation for exponential depreciation is

$$V = V(0) * K^{Y} = 31 * 0.86^{Y}.$$

Figure 10.1 shows its graph too. Notice that the line starts out above the curve. They meet again when $Y = 6.5$; thereafter the curve is above the line. In the linear model the machine changes value at a constant dollar rate: −2.9 thousand dollars per year. It is worthless when $V = 0$, when the stamper is 10.6 years old. In the exponential model the annual proportional change remains constant, so the dollar rate decreases. Each year it loses 14% of its value. In principle it is never worthless, though neither model is reasonable when Y is very large.

EXAMPLE 3. CALCULATOR SALES

The use of hand-held calculators has been growing at an increasing rate since their introduction early in the 1970s. One calculator company's sales increased from 1100 per month to 1700 per month in the 9-month period from January to September of 1973. Suppose the growth was exponential. Find the growth factor and an exponential equation for calculator sales. What were the sales in June 1975? When did the sales volume reach 10 500 per month? How long does it take sales to double?

First we choose names and units for the variables. Let C stand for monthly calculator sales, which depend on time T, which we measure in years Y, starting in January of 1973. Then the given data can be written:

$$C(0) = 1100, \quad \text{and} \quad C\left(\frac{9}{12}\right) = 1700,$$

since 9 months is 9/12, or 0.75 year. Thus the 0.75-year growth factor is 1700/1100. Hence the annual growth factor K is given by

$$K = \left(\frac{1700}{1100}\right)^{1/0.75} = 1.79,$$

and the exponential formula for the monthly sales is

$$C = 1100 * 1.79^{Y}. \tag{10.10}$$

Figure 10.2 shows the graph of this growth in sales.

To find the sales in June of 1975 we first note that June 1975 is 2 years and 5 months, or 2.42 years, from January 1973. Then

$$\begin{aligned} C(\text{in June 1975}) &= C(2.42) \\ &= 1100 * 1.79^{2.42} \\ &= 4472 \\ &\sim 4500. \end{aligned}$$

Next we are to find out when calculator sales reached 10 500 per month. We substitute $C = 10\,500$ in Equation 10.10 to obtain

$$10\,500 = 1100 * 1.79^{Y}, \tag{10.11}$$

and try to solve for Y. We have never encountered a problem like this, where the unknown appears as an exponent. We begin by dividing each side of Equation 10.11 by 1100, giving

$$1.79^{Y} = 9.55. \tag{10.12}$$

But we are stuck! In the next chapter we will learn a method for attacking Equation 10.12. All we can do now is try to guess the answer, and check our guess. The graph in Figure 10.2 suggests a guess of $Y = 4$.

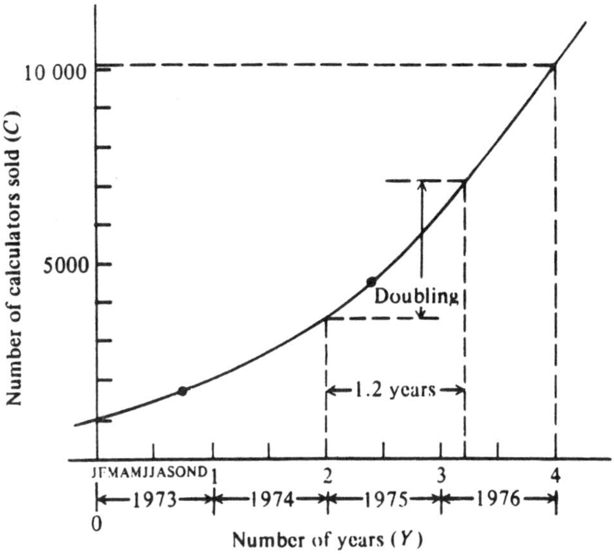

Figure 10.2 Monthly calculator sales

We calculate this (using the more precise annual growth factor of 1.786 796 . . ., saved all along in the calculator's memory):

$$1.79^{4} = 10.19$$

and compare the answer with the right side of Equation 10.12. We see that $Y = 4$ is too large. So we guess again: $Y = 3.5$. Then

$$1.79^{3.5} = 7.63,$$

which is too small. We try again, choosing Y between 3.5 and 4, but nearer to 4:

$$1.79^{3.8} = 9.07.$$

This is still too small. And so we try 3.9:

$$1.79^{3.9} = 9.62.$$

This is a little too large, but close enough for the number of significant digits given in the data. The solution to Equation 10.12 is a value of Y lying between 3.8 and 3.9 years, probably nearer the latter. Since 3.9 years is about 3 years and 11 months, calculator sales reached 10 500 per month in December of 1976.

Finally, let us find the time it takes for calculator sales to double. Doubled sales mean the growth factor is 2. That is, we are looking for the value of Y for which

$$1.79^Y = 2.$$

Here too the only method we know now to find Y is by trial and error. Try $Y = 2$:

$$1.79^2 = 3.19,$$

so in two years monthly sales more than triple. They double sometime between $Y = 1$ and $Y = 2$. Try $Y = 1.5$:

$$1.79^{1.5} = 2.39.$$

This is too big. Try $Y = 1.3$:

$$1.79^{1.3} = 2.13.$$

Still too big. So try $Y = 1.2$:

$$1.79^{1.2} = 2.01.$$

That's close enough. The doubling time is 1.2 years, or one year and two and a half months. This is shown in Figure 10.2. Sales double in any 1.2-year period–for example between January, 1975 and March, 1976.

PROBLEMS

*1. A Holbein sketch sold at auction for \$53 000 in 1950 and for \$86 000 four and one half years later. Assume the growth in value was exponential. Find a formula showing how the value depends on time. Graph this dependency. What is the sketch worth now? What was it worth in 1948? Estimate from your graph how long it takes to double in value, and when it was worth \$210 000.

2. The population of a small country was 2.3 million in 1965 and 3.4 million 3 years later.
   - (a) Find the yearly growth factor. Write an exponential equation showing how the population depends on the time T in years since 1965.
   - (b) Assuming the growth continues exponentially, find the population in 1960, 1970, 1973, and 1979.

(c) Draw a graph illustrating this growth.

(d) When will the population reach 15 million? When was it 2 million?

(e) Write a formula expressing the population when the time T is measured in years since 1960.

3. A truck was worth $12 500 when it was a year old and $7300 at age 3 years 5 months.

(a) Assuming the depreciation is exponential, find a formula with which to compute the value in terms of the age. Draw the graph. What is the value of the truck at age 10 years?

(b) For old times' sake, answer the same questions under the assumption that the depreciation is linear. Draw your graphs on the same set of coordinate axes.

*4. A colony of 16 500 bacteria grew to 21 500 in 2 hours 20 minutes. Find an exponential formula that describes this growth. Draw the graph. Estimate the time it takes for the colony to reach 26 500.

5. According to the Regional Director of the Census Bureau for New England, the world population was estimated to be 4 469 934 000 in 1980 and is projected to be 6 350 258 000 in the year 2000.

(a) What annual growth factor and annual percentage rate of growth were used to make the prediction?

(b) How do these figures compare with the rough estimates used in Problem 11 of Chapter 9?

*6. Suppose the annual inflation rate is 12%.

(a) What is the monthly rate?

(b) How long will it take prices to double? (Use your calculator to find, by trial and error, the time period that gives a growth factor of 2.)

7. If the inflation rate is 12% this year and was 18% last year what is the average rate for the two-year period?

8. If the inflation rate is 8% for nine months, then 10% for three months, what is the inflation rate for the year?

9. Suppose that inflation is doubling prices every 8.5 years. Find the annual growth factor and the annual inflation rate. (*Hint*: What is the 8.5-year growth factor?)

10. (a) Use the data in the first paragraph of Problem 14 of Chapter 9 to find a formula of the form

$$M = \text{intercept} * (\text{growth factor})^T$$

expressing the dependence of the metabolic rate M on the Celsius temperature T.

(b) Compare your answer with the formula given in part (c) of that problem. You should get the intercept shown there if you use the 1° growth factor 1.072, and the intercept 300 Cal/day if you use the growth factor $2^{(1/10)}$ without rounding.

*11. The population of Mexico tripled between 1945 and 1980 and is expected to double in the period from 1980 to 2000.

(a) Find the average annual percentage of increase in the population of Mexico for each of these two time intervals.

(b) Find the average annual percentage of increase for the entire 55-year period.

12. Inventory costs for the firm of White and Daughters were \$21 000 and \$43 000 in 1967 and 1977 respectively.

(a) Discuss the growth of these costs under each of the following assumptions: linear growth, and exponential growth. Draw graphs, on the same set of axes. Write equations. Use each model to predict 1981 inventory costs.

(b) Suppose you discover that in 1972 inventory costs were \$29 000. Which prediction for 1981 becomes more likely?

13. Air pressure decreases exponentially with altitude. At sea level in Boston (altitude 0 feet), the pressure is 30 inches of mercury. In Denver, which is a mile high, the pressure is 25.3 inches of mercury.

(a) Write a formula that relates air pressure (in inches of mercury) to altitude (in feet above sea level).

(b) What will the pressure be at the top of Mt. Everest (altitude 29 002 ft)? At the bottom of Death Valley (300 ft below sea level)?

(c) Draw a graph showing this information.

(d) The Weather Almanac says that a reasonable linear approximation is that pressure drops one inch for each 1000-foot increase in altitude. Write the corresponding linear formula. Identify the slope and intercept, with their units. Draw a linear graph on the same set of axes as your answer to (c).

(e) What is the linear prediction for the top of Mt. Everest?

(f) For what range of altitudes does the linear approximation give pressures that differ from the correct pressure by less than 0.1 inch?

14. Sales of trendy, fashionable items grow exponentially at first. If Wing Aero sold 140 softer saucers in the first week of September and 360 in the third week, write a formula showing how saucer sales depend on time (measured in weeks since the first week of September). What will sales be the week before Christmas?

*15. Stock bought in 1930 for $1000 was sold in 1978 for $46 500. What was the annual growth rate for the value of the stock?

16. The cost of heating a house depends exponentially on the difference between the inside and outside temperatures. Suppose that it costs Walt Reis $10 to keep the temperature inside at 66°F on a day when it's 30°F outside, and $13 when it's 22°F outside. What will it cost when the outside temperature drops to 5°F? How much can he save on such a day by setting his thermostat inside at 58°F?

17. Each of the following calculations can be done with or without the [y^x] key. For example, the answer to (a) can be found with either of the sequences

1.7 [y^x] 1 [+/−] [=],

or, as indicated in the second column,

1.7 [$1/x$].

In each case carry out the calculation both ways. Compare the answers and the speed of calculation.

(a) 1.7^{-1}; $\frac{1}{1.7}$ using [$1/x$] key
(b) $1.7^{0.5}$; 1.7 using square root key
(c) 1.7^2; 1.7 * 1.7
(d) 1.7^2; 1.7 squared using [x^2] key
(e) 1.7^1; 1.7 (no calculation to do)
(f) 1.7^0; 1 (no calculation to do)

18. (a) Can you calculate 1.7 * 1.7 * 1.7 in the time it takes the calculator to find 1.7^3 using the [y^x] key?
(b) Which is a faster way to calculate 1.7^4:

1.7 [y^x] 4 [=] or 1.7 [x^2] [x^2]?

Why does the second method work? What power do you raise a number to if you press the squaring key three times?

*19. (a) What should be the result of calculating $(2^{0.1})^{10}$?

(b) What is the result of that calculation? If it is not what you expected, try to explain why it's not.

CHAPTER 11

Logarithms

What's in a name? That which we call a rose by any other word would smell as sweet.

– *Romeo and Juliet*, Act II, Scene 2

Calculating with and understanding the meaning of logarithms
Simple transcendental equations

EXAMPLE 1. AN ARTIFICIAL PROBLEM

In a recent sci-fi horror story, an epidemic spreads with a daily growth factor of 10. For what time period is the growth factor 75? For what time period is it 600?

The example is truly artificial. The numbers are as extreme as the science-fiction background. A daily growth factor of 10 means that if one person is infected on Sunday, 10 will be on Monday, 100 on Tuesday, 1000 on Wednesday, and 10^6 or 1 million on Saturday. Therein lies the clue to the variables in the problem. We wish to study how the T-day growth factor, call it Q, depends on T. We know from Equation 9.10 that when the daily growth factor is K, the T-day growth factor is K^T. Thus in this example

$$T\text{-day growth factor} = Q = 10^T. \tag{11.1}$$

Figure 11.1 shows the graph of this dependency.

We have chosen to study a problem based on the outrageously large daily growth factor of 10 because powers of 10 are easier to understand than, say, powers of 1.7. By the end of the chapter we will be able to deal with any daily growth factor. But now the question asked is: When the daily growth factor is 10, for what time period T is the growth factor 75? We substitute $Q = 75$ in Equation 11.1 and try to solve the resulting equation,

$$10^T = 75, \tag{11.2}$$

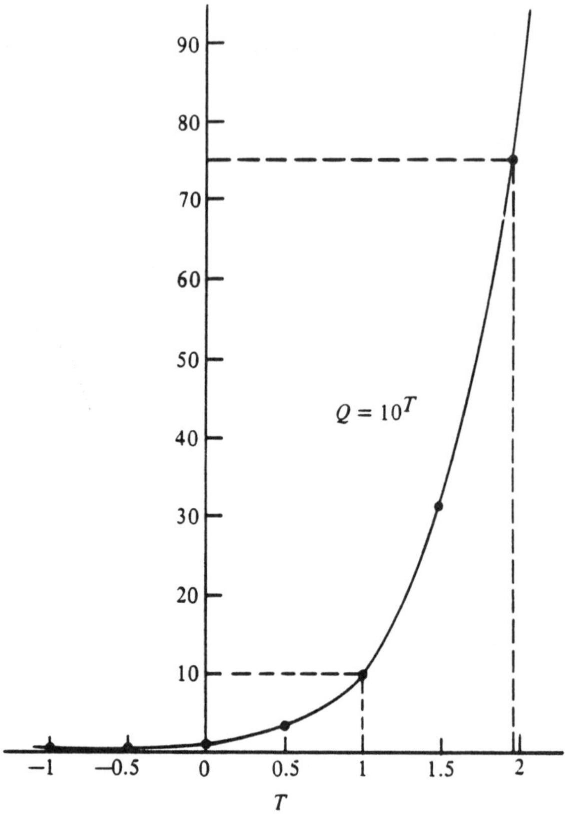

Figure 11.1 Rapid exponential growth

for T. We solved a problem like this in Example 4 of Chapter 10, by trial and error. Here the daily growth factor is 10 and the 2-day growth factor is $10^2 = 1000$, so the solution T is somewhere between 1 and 2 days. Figure 11.1 suggests that T is nearly 2, and so we start by guessing $T = 1.9$. A calculator check shows that

$$10^{1.9} = 79.43.$$

This is close to 75 but too large. When $T = 1.8$,

$$10^{1.8} = 63.10.$$

That's too small and not as close as when $T = 1.9$. Thus the answer is between 1.8 and 1.9, and nearer 1.9. The guess 1.87 yields

$$10^{1.87} = 74.13,$$

while 1.88 gives

$$10^{1.88} = 75.86.$$

Thus the answer is between 1.87 and 1.88. We could continue this guessing as long as we had decimal places on our calculator to play with. The next sequences of guesses would tell us that T is between 1.875 and 1.876, then between 1.8750 and 1.8751.

There must be a better way–and there is. Your calculator knows something about successive approximations: that is how it finds y^x. And it has another key, labeled [LOG], devoted precisely to the approximations we are using to solve Equation 11.2. If you enter

75 [LOG]

the calculator will pause for a while and then display the number

$$1.875\ 061\ 263 \qquad (11.3)$$

or some part thereof, depending on how many digits it can handle. This is the number we have been aiming for, between 1.8750 and 1.8751. Check that it really solves Equation 11.2 by calculating

$$10^{1.875\ 061\ 263} = 75.000\ 000\ 000.$$

The number in line (11.3) is called the *logarithm* of 75, or log 75, or log(75). Thus we write

$$\log 75 = 1.875\ 061\ 26\ldots. \qquad (11.4)$$

Notice that in Equation 11.4 the word *log* precedes the 75 although when we calculate we enter the number 75 before pressing the [LOG] key.

The logarithm key works similar magic for any value of Q in Equation 11.1. The second question in this example asks for the time T which gives a growth factor of $Q = 600$. Since $Q = 100$ when $T = 2$, and $Q = 1000$ when $T = 3$, the answer is between 2 and 3. But we need not think even that much to find it. Just enter

600 [LOG]

and read the result:

$$\log 600 = 2.7782.$$

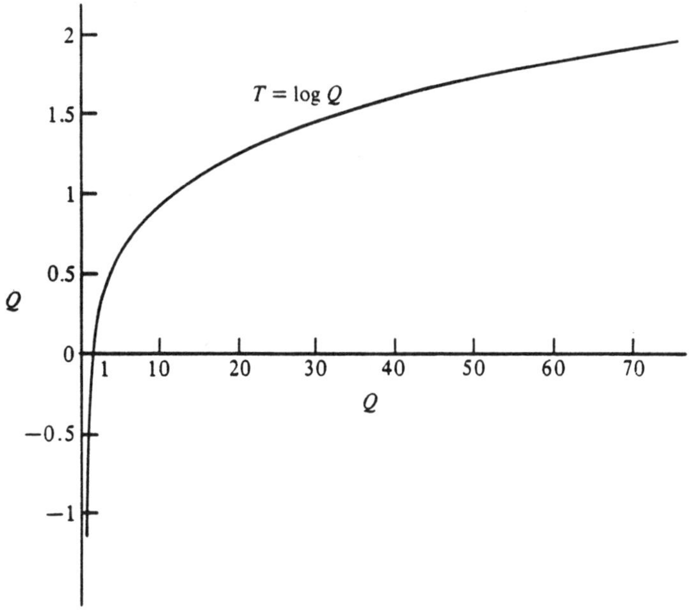

Figure 11.2 The logarithm curve

To check your answer, calculate

$$10^{2.7782}$$

and see that the answer is 600. The curve in Figure 11.2 shows how log(Q) depends on Q. The shape is the same as that of the curve in Figure 11.1, with the axes reversed.

EXAMPLE 2. WHO IS BURIED IN GRANT'S TOMB?

Grant, of course. The answer is part of the question.

Here's another riddle. What is the value of

$$10^{(\log 63)}?$$

You do not need a calculator to know that the answer is 63, because the answer is part of the question. Just think about what logarithms do. We discovered in Example 1 that calculating log 63 is the way to solve

$$10^T = 63,$$

and so substituting log 63 for T in that equation is just checking the solution.

Since we did not use the calculator to answer the riddle, we can answer it when it contains a variable instead of the number 63. We can understand logarithms as well as calculate with them:

$$\log Q \quad \text{is the name we give to the solution to} \quad 10^{?} = Q.$$

Therefore it is automatically always true that

$$10^{\log(Q)} = Q, \tag{11.5}$$

even when we don't know Q. Equation 11.5 is true for the same kind of reason that it is Grant who is buried in Grant's tomb. That's how the tomb got its name. Another useful way to think about the meaning of logarithms is to see that the equations

$$Q = 10^T \tag{11.6}$$

and

$$T = \log Q \tag{11.7}$$

each say exactly the same thing. Logarithms are just what we need to solve Equation 11.6 for T in terms of Q.

Logarithms are like square roots in that each was invented to solve an equation. Perhaps pursuing that analogy will help demystify logarithms. Without square roots, solving the equation

$$G = X^2 \tag{11.8}$$

for X in terms of G can be done only by trial and error. But with them, you need only say that

$$X = \sqrt{G}. \tag{11.9}$$

Equations 11.8 and 11.9, like Equations 11.6 and 11.7, say the same thing in two ways. The analog of Equation 11.5 when discussing square roots is

$$(\sqrt{G})^2 = G, \tag{11.10}$$

which is automatically true, even when we don't know G. If the analogy between logarithms and square roots—that each is used to name the solution to an equation—helps you understand logarithms, good. If not, just concentrate on the logarithms.

What is log(100)? We need no calculator to answer. Recall that log(100) names the solution to

$$10^{?} = 100. \tag{11.11}$$

But we know that $10^2 = 100$, so the solution to Equation 11.11 is 2. Thus

$$\log(100) = \log(10^2) = 2.$$

Similarly, since 6 solves the equation

$$10^{?} = 10^6,$$

we know that

$$\log(10^6) = 6.$$

In general, whatever value T has,

$$\log(10^T) = T, \tag{11.12}$$

because the solution to the equation

$$10^{?} = 10^T$$

is just T.

Equation 11.12 is a companion to Equation 11.5. Each is true because of the meaning of *logarithm* rather than for numerical reasons. The analog of Equation 11.12 for square roots is

$$\sqrt{X^2} = X. \tag{11.13}$$

The operations "raise 10 to the power ___" and "find the logarithm of ___" are intimately related. Each undoes the other. If you start with any number T, find 10^T, and then find the logarithm of the result, you are back at T where you began. This is Equation 11.12. If you start with any positive number Q, find $\log(Q)$ and then raise 10 to that power you are back at Q where you began. This is Equation 11.5. The same relation holds between the operations "square ___" and "find the square root of ___", and between the operations "double ___" and "find half of ___". Each undoes or inverts the work of the other.

Calculator designers often keep these special relationships in mind. On some calculators, when you use the [LOG] key after you have pressed a key labeled [INV] or [2ND] the operation performed is "10 to the power ___." Thus the instructions

75 [LOG]

produce

$$\log 75 = 1.875\ 061\ 263,$$

while the instructions

1.875 061 263 [INV] [LOG]

produce

$$10^{1.875\ 061\ 263} = 75.000\ 000\ 000.$$

In particular, you can calculate 10^T without using the y^x key. Just invert the logarithm. If your calculator does not compute 10^x in this way then it probably has a separate [10^x] key, which you can use instead of [y^x].

If you are told that

$$\log x = 2.3,$$

and are asked to find x, you must invert the logarithm:

$$x = 10^{2.3}.$$

Although you may *calculate* that as

2.3 [INV] [LOG],

you should *think* of it as $10^{2.3}$. Saying "inv log x" when you mean "10^x" is like describing the number "twice x" as "the opposite of half of x." It is roundabout and clumsy.

You may also find that "square root of ___" and "square

____" are handled by a single key, with one of the operations called by first using [INV] or [2ND], because these operations too invert or undo one another.

Next we use our newly won understanding of logarithms to explore some difficult ideas. First, what is the value of log(0.01)? We know from our study of exponents that

$$0.01 = \frac{1}{100} = 10^{-2},$$

so it follows from Equation 11.12 that

$$\log(0.01) = \log(10^{-2}) = -2. \tag{11.14}$$

Similarly,

$$\log(0.1) = \log(10^{-1}) = -1, \tag{11.15}$$

and, finally (and sometimes confusingly)

$$\log(1) = \log(10^{0}) = 0. \tag{11.16}$$

These facts all show on the graph of the equation $T = \log(Q)$ in Figure 11.2. We can understand them better if we think again about growth factors. Recall that in Example 1 we thought of $Q = 10^T$ as the T-year growth factor. If we are given the growth factor Q, and want to find the time it takes to achieve that growth factor, we just compute $\log(Q)$.

Suppose $Q = 1$. How long does it take until the growth factor is 1? This growth factor corresponds to no change at all, which takes no time at all: 0 days. That is why $\log(1) = 0$. Suppose next that the time is negative, so we are looking into the past. Then the growth factors are less than 1 (but still positive). Thus when $0 < Q < 1$, we expect $\log Q$ to be a negative time. That is confirmed by Equations 11.14 and 11.15.

Thus the result of calculating a logarithm is *negative* when the number we start with is *less than 1*. What if we start with a negative number and try to compute its logarithm? For example, what is log(−3)? The calculator gives an error message. To see why, think about what log(−3) means. It's the name we give to the solution to the equation

$$10^{?} = -3.$$

But this equation has no solution. There is no such thing as a negative growth factor. No power of 10 gives −3 for an answer. Thus *you can-*

Table 11.1 Some logarithms, and properties of logarithms

Q	$\log Q$
100	2
75	1.875 . . .
10	1
1	0
0 1	−1
0.01	−2
10^T	T
1	0
more than 1	positive
less than 1	negative
0 or negative	no such thing

not find the logarithm of a negative number, just as *you cannot find the square root of a negative number*.

The discoveries we have been led to by exploring the riddles "Who is buried in Grant's tomb?" and "What is $10^{\log(Q)}$?" are summarized symbolically in Table 11.1. Parts of this table echo parts of Table 9.3. In the rows telling when logarithms are positive and negative, the logarithm acts as a dictionary for translating the exponential idea of a growth factor into the linear idea of a slope. We shall pursue this theme further in this chapter and the next.

EXAMPLE 3. THE LOGARITHM OF A PRODUCT

If in Example 1, the growth factor for A days is 75, and the growth factor for B days is 63, what is the growth factor for $A + B$ days?

We can answer this question without knowing about logarithms and without knowing values for A and B. If we wait first for A days and then for B days, the overall growth factor for the $(A + B)$-day period will be the *product* of the two growth factors, 75 * 63. To see this fact algebraically, remember that

$$10^A = 75 \quad \text{and} \quad 10^B = 63, \tag{11.17}$$

Because 75 and 63 are the A- and B-day growth factors. Then the $(A + B)$-day growth factor is

$$10^{A+B} = 10^A * 10^B = 75 * 63. \tag{11.18}$$

Now on to logarithms. Equations 11.17 say that

$$\log 75 = A \quad \text{and} \quad \log 63 = B \tag{11.19}$$

while Equation 11.18 is equivalent to

$$\log(75 * 63) = A + B. \tag{11.20}$$

When we combine Equations 11.19 and 11.20 we get

$$\log(75 * 63) = \log 75 + \log 63. \tag{11.21}$$

A and B do not appear in Equation 11.21, and nothing in the argument that led to it treated the numbers 75 and 63 in any special way. Therefore, for any numbers Q and R,

$$\log(Q * R) = \log Q + \log R. \tag{11.22}$$

That is:

Logarithms turn products into sums.

This is the central, most useful property of logarithms. It is why they were invented. Addition is easier than multiplication, and so in the days before calculators anything that made the former from the latter was a valuable tool. Now our machines multiply almost as easily as they add, so that logarithms are rarely used for difficult arithmetic. But they have other uses. One we have already met: They find the time T in terms of the growth factor Q in Example 1.

The fact that logarithms turn products into sums makes them particularly compatible with scientific notation. For example, suppose we have computed

$$\log(2.7183) = 0.4343. \tag{11.23}$$

That the answer here is between 0 and 1 is no surprise, since 2.7183 is between 10^0 and 10^1. But what is $\log(2.7183 * 10^9)$? We need not use our calculator a second time, since Equation 11.22 tells us that

$$\begin{aligned} \log(2.7183 * 10^9) &= \log(2.7183) + \log(10^9) \\ &= 0.4343 + 9 \\ &= 9.4343. \end{aligned}$$

Similarly,

$$\begin{aligned}\log(2.7183 * 10^{126}) &= \log(2.7183) + \log(10^{126}) \\ &= 0.4343 + 126 \\ &= 126.4343.\end{aligned} \tag{11.24}$$

Notice that this is beyond the capability of your calculator, which cannot handle numbers bigger than 10^{100}. The calculation exemplified by Equation 11.24 shows that when

$$Q = m * 10^E$$

is written in scientific notation, so that m is between 1 and 10, then

$$\log(Q) \text{ is between } E \text{ and } E + 1. \tag{11.25}$$

Thus logarithms measure the *order of magnitude* of large numbers like those we encountered in Chapter 8. They also measure the order of magnitude of small numbers, whose logarithms are negative.

For example,

$$\begin{aligned}\log(0.0027183) &= \log(2.7183 * 10^{-3}) \\ &= \log(2.7183) + \log(10^{-3}) \\ &= 0.4343 + (-3) \\ &= -2.5657.\end{aligned}$$

The answer is negative, as we expected it to be. It is between -3 and $(-3 + 1) = -2$, which agrees with the estimate in statement (11.25).

Because negative logarithms often cause confusion we will consider them once more. What is $\log(1/2.7183)$? Since $1/2.7183$ is less than 1, its logarithm will be negative. Our calculators tell us that

$$\log\left(\frac{1}{2.7183}\right) = -0.4343. \tag{11.26}$$

The relationship between the numbers in Equations 11.23 and 11.26 is too striking to be accidental. It is an example of the general principle

$$\log\left(\frac{1}{Q}\right) = -\log(Q). \tag{11.27}$$

This reinforces our theme that logarithms translate exponential ideas into linear ones. Here division becomes subtraction. To see why Equation 11.27 is true, we use the fact that logarithms turn products into sums, in reverse. The sum of two logarithms is the logarithm of the product:

$$\begin{aligned}\log(Q) + \log\left(\frac{1}{Q}\right) &= \log\left(Q * \frac{1}{Q}\right) \\ &= \log(1) = 0.\end{aligned} \tag{11.28}$$

Equation 11.27 follows when we subtract the number $\log(Q)$ from each side of Equation 11.28.

EXAMPLE 4. A MILD EPIDEMIC

A flu (short for influenza) epidemic is spreading through Boston with a weekly growth factor of 1.3. For what time period is the growth factor 3?

This example resembles Example 1, but the numbers are more reasonable here. The growth factor for a T-week period is

$$G = 1.3^T. \tag{11.29}$$

The problem asks us to substitute $G = 3$ in Equation 11.29 and solve the resulting equation,

$$1.3^T = 3, \tag{11.30}$$

for T. This brings us back to where we started in Example 1. We can try guessing–the answer must be between 4 and 5, since $1.3^4 = 2.9$ and $1.3^5 = 3.7$. Or we can hope to find a logarithm key on our calculator that inverts 1.3^T in the same way [LOG] inverts 10^T. Better yet, we can learn a little more about logarithms. The little more we need is the fact that logarithms turn exponentiation into multiplication:

$$\log(K^T) = T * \log(K). \tag{11.31}$$

We shall use Equation 11.31 to solve the problem, and then see why it is true.

Calculate the logarithm of each side of Equation 11.30:

$$\log(1.3^T) = \log 3.$$

Then, using Equation 11.31, we find that

$$T * \log 1.3 = \log 3,$$

and thus

$$\begin{aligned} T &= \frac{\log 3}{\log 1.3} \\ &= \frac{0.4771}{0.1139} \\ &\sim 4.2 \text{ weeks.} \end{aligned} \tag{11.32}$$

That is the time it takes for the number of infected Bostonians to triple. [*Warning*: To calculate the quotient log(3)/log(1.3) in Equation 11.31, you must first find the logarithms separately and then divide. This is because log(3/1.3) is different, as is log(3 − 1.3).]

To see why Equation 11.31 is true, begin again with the meaning of logarithms. We know that $\log(K^T)$ solves the equation

$$10^? = K^T. \tag{11.33}$$

If we can show that $T * \log K$ also solves this equation, then $\log(K^T)$ and $T * \log K$ will have to be the same number. But this isn't too difficult. Use the proper rule for exponents:

$$\begin{aligned} 10^{(T * \log K)} &= 10^{[(\log K) * T]} \\ &= (10^{\log K})^T = K^T. \end{aligned}$$

Done!

EXAMPLE 5. THE PRICE OF A BICYCLE

If the annual inflation rate is 11.3%, how long will it be until a bicycle that now costs $150 costs $250?

Knowing what we now know, this question is straightforward. The variables are B, the dollar cost of the bicycle, and Y, the time in years from now. The annual growth factor is $K = 1.113$, and the equation showing how B depends on Y is

$$B = 150 * 1.113^Y. \tag{11.34}$$

Now substitute $B = 250$ in Equation 11.34 to get:

$$250 = 150 * 1.113^Y, \tag{11.35}$$

and solve for Y. Begin by calculating logarithms.

$$\begin{aligned} \log 250 &= \log(150 * 1.113^Y) \\ &= \log 150 + \log(1.113^Y) \\ &= \log 150 + Y * \log(1.113). \end{aligned} \tag{11.36}$$

All these logarithms may look nasty, but they are just numbers. The equation is simply linear in the variable Y. Since $\log(250) = 2.40$, $\log(150) = 2.18$, and $\log(1.113) = 0.046$, Equation 11.36 says

$$2.40 = 2.18 + Y * 0.046.$$

Thus

$$Y * 0.046 = 2.40 - 2.18$$

and

$$Y = \frac{2.40 - 2.18}{0.046} = 4.8 \text{ years.}$$

Thus the bicycle will cost \$250 in 4 years and 9 months.

Here are two other correct ways to solve the problem. Each uses logarithms to bring the unknown Y down from the exponent onto the line

We might delay taking logarithms in Equation 11.35 by first dividing each side by 150, giving

$$\frac{250}{150} = 1.113^Y.$$

Then

$$\log \frac{250}{150} = Y * \log(1.113)$$

so

$$Y = \frac{\log(250/150)}{\log(1.113)}$$

$$= \frac{\log(1.67)}{\log(1.113)}$$

$$= \frac{0.22}{0.46}$$

$$= 4.8.$$

Or we could solve Equation 11.34 for Y in terms of B before substituting $B = 250$. Taking logarithms:

$$\begin{aligned} \log B &= \log(150 * 1.113^Y) \\ &= \log(150) + \log(1.113^Y) \\ &= \log(150) + Y * \log(1.113), \end{aligned}$$

so

$$Y * \log(1.113) = \log B - \log 150$$

and hence

$$Y = \frac{\log B - \log 150}{\log 1.113}$$

$$= \frac{\log B - 2.18}{0.0465}$$

Then we need only substitute $B = 250$, or any other value we are interested in.

PROBLEMS

1. Estimate the value of each of the following logarithms. Check the estimates with your calculator.
   (a) $\log(2)$ (b) $\log(81)$
   (c) $\log(164.7)$ (d) $\log(1\ 000\ 000)$
   (e) $\log(100.001)$ (f) $\log(0.1234)$
   (g) $\log(6 * 10^{23})$ (h) $\log(6 * 10^{-23})$
   (i) $\log(6 * 10^{123})$ (j) $\log(6 * 10^{-123})$

*2. (a) Solve for t: $19.6 = 4.8 * 1.7^t$
   (b) Solve for u: $3.1 = 4.8 * 1.7^u$
   (c) Solve for r: $19.6 = 4.8 * 0.83^r$
   (d) Solve for s: $3.1 = 4.8 * 0.83^s$

3. Evaluate the following:
   (a) $10^{\log(234.567)}$
   (b) $\log(10^{-4.321})$
   (c) $\log[\log(1000)]$
   (d) $10^{(10^{0.477\,121})}$
   (e) $A - \log(10^{1+A})$

4. What color was George Washington's white horse? Why is this question included here?

*5. Evaluate the following:
   (a) $3 + \log 2$
   (b) $\log 3 + 2$
   (c) $\log(3 + 2)$
   (d) $\log\sqrt{3 + 2}$
   (e) $\log(3 + \sqrt{2})$
   (f) $3 + \log\sqrt{2}$

6. Solve for A: $3 + \log(A) = 1.2$
   Solve for B: $\log 3 + B = 1.2$
   Solve for C: $\log(3 + C) = 1.2$
   Solve for D: $\log\sqrt{3 + D} = 1.2$
   Solve for E: $\log(3 + \sqrt{E}) = 1.2$
   Solve for F: $3 + \log\sqrt{F} = 1.2$

*7. Solve for A: $\log(A) + \log(2A) = 5$
   Solve for B: $\log(B) + \log(B^2) = 5$
   Solve for C: $\log(C) + 2\log(C) = 5$
   Solve for D: $10^{(\log 5D)} = D + 8.$
   Solve for E: $\log(E) = -234.56$
   Solve for F: $F = \log(10^{147.2})$
   Solve for G: $\log(3.61 + 1.73\,G) = 2.92$
   Solve for H: $\log(5 + H) = 0.97$
   Solve for K: $10^{48K} = 163$
   Solve for L: $10^{(48 + L)} = 163$

8. (a) Suppose that $K^{1.21} = 7.63$.
   Find K in two ways: (1) by the method introduced in Example 1 of Chapter 10, and (2) by using logarithms.
   (b) Solve: $K^A = B$ for K in terms of A and B, in the same two ways.

*9. Solve: $P = P(0)\,K^t$ for K in terms of P, $P(0)$, and t;
   Solve for t in terms of P, $P(0)$, and K.

10. Solve: $R = 1 - K^D$ for D in terms of R and K.

11. When will the number of cars and trucks registered in the United States reach 220 million? (See Problem 13 of Chapter 9.)

12. When was the Holbein sketch that is the subject of Problem 1 in Chapter 10 worth \$210 000? When was it worth \$40 000? When was it (or will it be) worth \$1 million?

13. Use the facts that log 2 $\sim$ 0.30 and log 3 $\sim$ 0.48, the rules for logarithms, and *no calculator* to evaluate (approximately):
    (a) log 6 (b) log 4 (c) log 8 (d) log 9
    (e) log 5 (*Hint*: 5 = 10/2. The fact that 5 = 2 + 3 is of no help!)
    (f) log 10 (easy) (g) log 12
    (h) log 11 (*Hint*: 11 $\sim$ 100/9) (i) log 7 (*Hint*: 49 $\sim$ 50)
    Check your answers with your calculator.

14. (a) Show that the number half way between log(A) and log(B) is the logarithm of $\sqrt{AB}$.
    (b) Experiment with some numbers to see if you can discover where between log(A) and log(B) you will find the cube root of $A * B^2$. Start with $A = 10$ and $B = 1000$.

CHAPTER 12

Using Logarithms

Double, double, toil and trouble.
– *Macbeth*, Act IV, Scene I

The logarithms of exponentially growing quantities
Implicitly exponential scales / Doubling times and half-lives

EXAMPLE 1. ELECTRIC ENERGY CONSUMPTION

In 1970 the United States generated $1.6 * 10^{12}$ kilowatt-hours of electricity. Since 1920, electric energy use has increased at a rate of 7 percent per year. Find a formula showing how energy use depends on the time. Draw graphs illustrating how energy use and its logarithm depend on time.

The annual energy consumption P depends on T, measured in years since 1920, according to the formula

$$P = P(0) * K^T. \tag{12.1}$$

The annual growth factor K is 1.07, since P increases at 7% per year. To find the intercept $P(0)$, the energy consumption in 1920, we observe that 1920 is 50 years earlier than the year in which $P = 1.6 * 10^{12}$ kWh. Hence

$$\begin{aligned} P(0) &= 1.6 * 10^{12} * 1.07^{-50} \\ &= 5.4 * 10^{10} \text{ kWh.} \end{aligned} \tag{12.2}$$

Before we graph the relation between P and T, we will change units to absorb the factor of 10^{10}. A kilowatt-hour is 10^3 watt-hours. We are dealing here with 10^{10} kWh, or 10^{13} Wh, which may also be expressed as 10 Terawatt-hours. (Remember that *Tera* means $* 10^{12}$.) When power consumption P is measured in TWh/year we know that

$$P = P(0) * 1.07^T = 54 * 1.07^T. \qquad (12.3)$$

Check Equation 12.3 by calculating the consumption in 1970.

$$P(50) = 54 * 1.07^{50} = 1600 \text{ TWh}.$$

This is what we began with, and so our formula checks.

Next we must graph the dependence of P and of $\log P$ on T. To that end, we have calculated the data in Table 12.1. Figure 12.1 shows the graph of P in terms of T. It is hard to extract useful information from this graph, because the growth is so rapid. When we try to put the numbers 54 and 47 000 on the same vertical scale, the former will be essentially indistinguishable from 0. The curve is nearly flat when T is less than 40; it is nearly vertical when T is greater than 60. We can display the numbers much better by graphing $\log P$ as in Figure 12.2. The striking inference from this graph is that $\log P$ seems to depend *linearly* on T. Why? Look at Equation 12.3, and compute the logarithm of each side:

$$\begin{aligned} \log P &= \log(54 * 1.07^T) \\ &= \log 54 + T * \log(1.07) \qquad (12.4) \\ &= 1.7 + 0.029 * T. \end{aligned}$$

This equation shows that the graph in Figure 12.2 actually is a line. Its intercept is $1.7 = \log 54$, the logarithm of the intercept in Figure 12.1. Its slope is $0.029 = \log(1.07)$, the logarithm of the growth factor. Each year the value of P is multiplied by the growth factor 1.07. Since

Table 12.1 Electric energy generated in the United States

Year	T (years since 1920)	P(TWh)	$\log P$
1920	0	54	1.7
1950	30	410	2.6
1960	40	810	2.9
1970	50	1 600	3.2
1980	60	3 100	3.5
1990	70	6 200	3.8
2000	80	12 000	4.1
2020	100	47 000	4.7

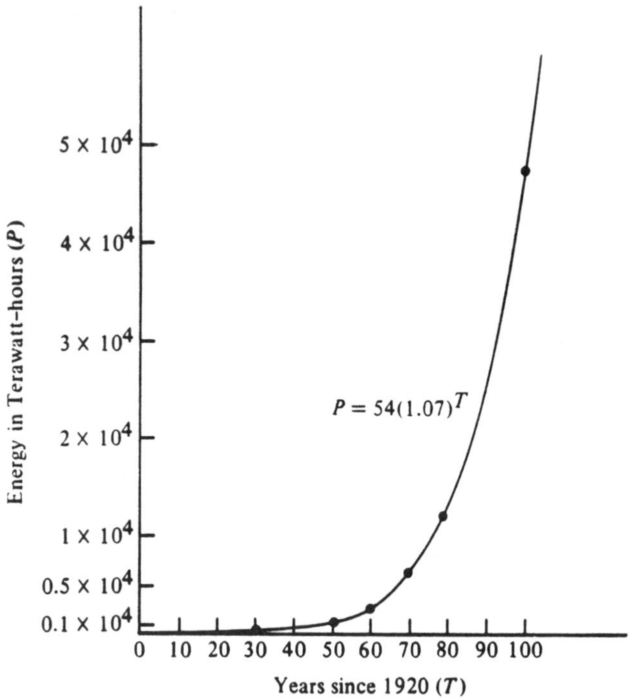

Figure 12.1 Electric energy consumed annually in the United States

logarithms turn * into +, each year we *add* to log P the logarithm of 1.07, which equals 0.029, the slope in Figure 12.2.

Each vertical change of 1 unit in Figure 12.2 adds 1 to log P. The effect of this is to multiply P by 10. To see this, we have drawn a scale for P parallel to the one showing log P. The values on the P scale are not evenly spaced. The distance between 10 and 20 is much larger than that between 90 and 100. It is the log P scale that is evenly spaced, so that equal spacing on the P scale corresponds to equal quotients rather than equal differences. That is why the distance between 50 and 100 is the same as the distance between 500 and 1000. This is yet another example in which logarithms serve as a dictionary for translating exponential ideas into linear ones. If P is any given quantity depending exponentially on another, T, so that

$$P = P(0) * K^T, \tag{12.5}$$

where $P(0)$ is the value of P when $T = 0$, and the growth factor is K, then $\log P$ depends linearly on T:

$$\log P = \log P(0) + (\log K) * T. \tag{12.6}$$

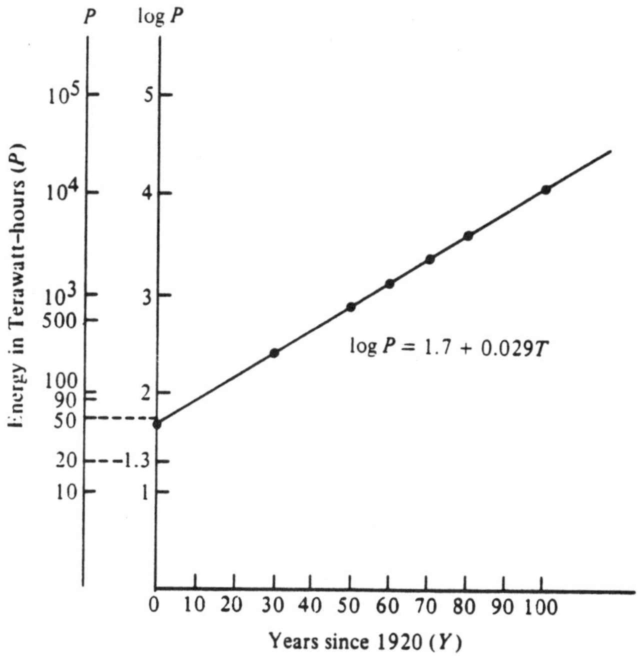

Figure 12.2 Electric energy consumed annually in the United States

The logarithm of the growth factor in Equation 12.5 is the slope in Equation 12.6.

EXAMPLE 2. THE pH SCALE

The water in New England lakes is more acidic than it was a generation ago, because the rain that falls is more acidic. In this example we discuss how acidity is measured.

The common unit for measuring acidity is called *pH*. Pure water has a pH of 7. (We shall see why soon.) Vinegar is quite acidic; its pH is 3. The lower the pH the more acidic the solution. A cleaning solution of household ammonia is less acidic than water; its pH is about 10. Let us see what chemical quantity pH really measures. Roughly speaking, a solution is acidic when it has free hydrogen ions (H^+) in it. These hydrogen ions can be put there by a chemist, or by natural processes. Even pure water contains some H^+ ions since water molecules, which consist of two atoms of hydrogen and one of oxygen, sometimes break up into two pieces, H^+ and OH^- ions. In pure water at room temperature there are about 10^{-7} grams of H^+ ions per liter and an equal number of OH^- ions. That is, the concentration of H^+ ions is 10^{-7} g/L. Chemists call the negative of the logarithm of the H^+-ion concentration the pH of the solution. Thus, pure water has a pH of

$$-\log(10^{-7}) = 7.$$

Since the concentration of H^+ ions in vinegar is about 10^{-3} g/L, vinegar has a pH of 3. It is 10^4 or 10 000 *times* as acid as pure water. An acid with a pH of 1 would be 10^6 or 1 million times as rich in H^+ ions as pure water. Ammonia has a pH of 10; the H^+-ion concentration is 10^{-10} g/L, only 1/1000 that of water.

The logarithm in the definition of pH makes exponential changes in H^+-ion concentration measurable on a linear scale. The pH in acidic New England lakes is decreasing linearly with time. That change tells us that the concentration of H^+ ions is increasing exponentially with time. And this is because the sulfur in oil and coal that are burned to drive cars and generate electricity produces sulfuric acid in rain water. We can infer from Example 1 that the use of fossil fuels is increasing exponentially with time.

There are other scales that, like the pH scale, are logarithmic. The *Richter scale*, which measures the strength of earthquakes, is one. An earthquake that measures 7 is 10 *times* as powerful as one that measures 6. An earthquake that measures 8 on the Richter scale is 10 000 times as powerful as one rated at 4. The *decibel* scale, which measures the loudness of sounds, is another logarithmic scale.

EXAMPLE 3. DOUBLING TIMES

If the rate of inflation is 12 percent per year, how long will it take for prices to double?

We can use logarithms to answer this question. The annual growth factor is 1.12; we seek the time T for which the growth factor is 2. That is, we must solve the equation

$$1.12^T = 2$$

for T. Taking the logarithm of each side gives

$$T * \log 1.12 = \log 2.$$

so

$$T = \frac{\log 2}{\log 1.12} = \frac{0.301}{0.0492} = 6.1 \text{ years.} \tag{12.7}$$

One of the most surprising, and important, facts about exponential growth is that the doubling time depends only on the growth factor–not on the initial value. In this example it will take 6.1 years for the price of $50 worth of groceries to double. It will take the same 6.1 years for the price of a $9000 automobile to double.

We can generalize the computation of the doubling time when $K = 1.07$ to find a formula for the doubling time D in terms of the growth factor K. D is the time period for which the growth factor is 2, and so

$$K^D = 2. \tag{12.8}$$

To solve for D, take logarithms:

$$D * \log K = \log 2.$$

Then

$$\begin{aligned}\text{doubling time} &= D \\ &= \frac{\log 2}{\log K} \\ &\sim \frac{0.3}{\log K}.\end{aligned} \tag{12.9}$$

In Equation 12.9 we used the approximation

$$\log 2 = 0.301\ 029\ 996 \sim 0.3.$$

That is an approximation worth remembering. Log 2 is very close to 0.3 because

$$2^{10} = 1024 \sim 1000 = 10^3. \tag{12.10}$$

Thus

$$10^3 \sim 2^{10}. \tag{12.11}$$

so

$$10^{0.3} = (10^3)^{0.1} \sim (2^{10})^{0.1} = 2. \qquad (12.12)$$

But Equation 12.12 says that log 2 ~ 0.3. That is to say, 10 doublings is roughly the same as multiplying by 1000.

Let us apply Equation 12.9 to find out how long it takes for annual electric energy consumption in the United States to double. Recall from Example 1 that the annual growth factor is 1.07. Thus the doubling time is

$$D = \frac{0.3}{\log 1.07} = \frac{0.3}{0.0294} = 10.24 \sim 10 \text{ years.}$$

This fact is visible in Table 12.1. Look at the rows for the years 1950–2000. The value of P just about doubles every 10 years. Every 10 years you add log 2 ~ 0.3 to log P.

Suppose a population of microorganisms grows exponentially with a doubling time D of 2.6 days. What is the daily growth factor K? We know that

$$D = \frac{0.3}{\log K}. \qquad (12.13)$$

We need only to solve this equation for K in terms of D and substitute $D = 2.6$. Trying to isolate K, we multiply each side of Equation 12.13 by log K and divide each side by D:

$$\log K = \frac{0.3}{D}. \qquad (12.14)$$

To undo the logarithm and find K, raise 10 to the power log K:

$$K = 10^{\log K} = 10^{0.3/D}. \qquad (12.15)$$

In this problem, $D = 2.6$, so that

$$K = 10^{0.3/2.6} = 1.3$$

is the growth factor. The microorganism population grows at the rate of 30% per day.

One last reworking of Equation 12.15 is instructive:

$$\begin{aligned} K &= 10^{0.3/D} \\ &= (10^{0.3})^{1/D} \\ &= 2^{1/D} \end{aligned} \tag{12.16}$$

But this follows directly from Equation 12.8. It is another way to find K when D is known.

EXAMPLE 4. HALF-LIVES

Find the half-life of iodine-131.

We must first learn what *half-life* means. It is to exponential decay what doubling time is to exponential growth. The half-life of iodine-131 (written ^{131}I) is the time it takes for half of what you have to decay. That time depends only on the growth factor, not on the amount of ^{131}I you start with. In Example 3 of Chapter 9 we saw that the daily growth factor for ^{131}I is $K = 0.9174$. The half-life is the time H for which

$$K^H = \frac{1}{2}. \tag{12.17}$$

Solving for H gives

$$H * \log K = \log\left(\frac{1}{2}\right) = -\log 2 \sim -0.3,$$

so

$$H \sim \frac{-0.3}{\log K}. \tag{12.18}$$

The half-life formula in Equation 12.18 resembles the doubling time formula in Equation 12.9. Only the sign has been changed. But notice that you compute a half-life only when the quantity is shrinking. Then the growth factor K is less than 1 so that its logarithm is negative. In this example, $K = 0.9714$ and

$$\begin{aligned} H &= \frac{-0.3}{\log(0.9174)} \\ &= \frac{-0.3}{-0.0374} \\ &= 8.013 \\ &\sim 8 \text{ days.} \end{aligned}$$

If you start with 1000 g you will have 500 g left after 8 days, 250 g after 16 days, 125 g after 24 days (three half-lives), and so on. After 80 days, or 10 halvings, you will have only about 1 g, or one one-thousandth of what you began with. This is the approximation we saw in Equation 12.10. Ten halvings corresponds roughly to dividing by 1000.

We can also express the growth factor K in terms of the half-life H by solving either of Equations 12.17 or 12.18 for K. The result is

$$K = \left(\frac{1}{2}\right)^{1/H} = 10^{-0.3/H}.$$

For example, if a panel stamping machine worth \$57 000 has a half-life of 5.5 years, then it is depreciating exponentially with a yearly growth factor of

$$K = 0.5^{1/5.5} = 0.88.$$

The rate of depreciation is just under 12% per year.

PROBLEMS

*1. The population of a small country increased from 2.3 million in 1970 to 3.7 million in 1975. The growth was exponential.
   (a) Find a formula and draw a graph showing how the population grew.
   (b) Find a formula and draw a graph showing how the logarithm of the population grew.
   (c) Predict the population in 1980, 1985, 2000, and 2025.
   (d) When will the population reach 10 million?
   (e) How long does it take to double?

2. Restudy the rabbit population explosion (Problem 11 of Chapter 9.) This time graph the logarithm of the number of rabbits.

3. (a) Use the data in Figure 1.7 to draw a graph showing how the logarithm of the United States population changed with time during the period between the Revolutionary and Civil Wars (1790–1860).

   (b) Draw the straight line that best fits these points. Estimate its slope, and use the result to write an exponential formula for the population growth.

   (c) Use the line you found in (b) to extrapolate to 1980. Is the population so predicted higher or lower than the actual 1980 population?

*4. Plutonium-238 (^{238}Pu) is useful in bombs and power plants, but is a dangerously radioactive chemical element that decays slowly into less dangerous elements. If you started now with 100 grams, in a year you would still have 99.2 grams.

   (a) Find a formula for computing the amount P of ^{238}Pu as a function of the number of years Y from now. How much ^{238}Pu will you have in the year 2000? In 2081? In 500 years?

   (b) Draw graphs showing how P and $\log(P)$ depend on Y.

   (c) What is the half-life of ^{238}Pu?

   (d) When will you have only 1 milligram of ^{238}Pu? (Even that much can be lethal!)

5. Geologists have recently recalculated the magnitude of the 1907 San Francisco earthquake. It was thought to have registered 8.25 on the Richter scale. Current opinion is that it was only of magnitude 7. How much less powerful was it according to the new estimate?

6. (a) Find the doubling times for annual inflation rates of 2%, 5%, 10%, 15%, and 20%.

   (b) What inflation rate will cause prices to double in 5 years? In 10 years? In 20 years?

*7. A population of 259 000 people has a doubling time of 13.4 years. Write an exponential formula for the growth. Draw a graph. When will the population reach 1 million?

8. The concentration of drugs in the bloodstream decreases exponentionally over time. For Gigamycin the half-life is 5.2 hours. At noon a patient receives an injection that brings his blood Gigamycin concentration to 6.5 μg/L (micrograms per liter). A second injection is to be administered when the concentration falls to 1.4 μg/L. When will that be?

*9. Hydrogen-3 (tritium, ^3H, an isotope of hydrogen used in fusion-power experiments) has a half-life of 12.5 years. Answer without using your calculator: How long does it take for 75% of a quantity of tritium to decay? Approximately how long will it be before there is one thousandth of the original quantity left?

10. A population of houseflies doubles every 16 days. If there are 10 000 today, how many will there be tomorrow? In a week?

11. Excerpt from the January, 1982 *Scientific American*, p. 77: "In the manufacture of microelectronic devices the area on a silicon chip required for [a single memory location] has been shrinking by a factor of two every 18 months." Suppose a silicon chip held 2000 memory locations in 1970.

(a) Write a formula showing how the number of memory locations the chip can hold varies with time. Draw a graph.

(b) How many locations can a similar chip hold now? If present trends continue, when will such a chip be able to carry one million memory locations?

*12. In every half century since 1750 the number of scientific periodicals has increased tenfold. By 1950 there were 100 000 journals. (*Science*, Number 4467.) Try to answer the following questions without using your calculator.

(a) How many journals were there in 1750?

(b) If the growth continues at the same rate, how many will there be in 2000? In 2050?

(c) What is the doubling time?

13. Willard Libby won a Nobel prize for devising a technique for dating archaeological specimens which exploits the fact that carbon-14 has a half-life of 5700 years. He observed that the ratio of radioactive ^{14}C to stable ^{12}C decays exponentially at the same rate as does ^{14}C.

(a) The $^{14}C:^{12}C$ ratio in an ancient wooden idol is found to be 60% of the naturally occurring ratio. How old is the idol? (*Hint*: Use the half-life figure to find the yearly shrinking factor for the $^{14}C:^{12}C$ ratio.)

(b) A similar technique is used to date rocks. There it is the ratio of ^{40}K (potassium-40) to ^{39}K which decays exponentially, with a half-life of $1.4 * 10^9$ years. The oldest known rocks have a $^{40}K:^{39}K$ ratio that is 18% of the natural ratio. How old are the rocks? (*Hint*: Measure time in billions of years, and use the given half-life to find the 1-billion-year shrinkage factor for the $^{40}K:^{39}K$ ratio. What happens if you try to find the year shrinkage factor?)

*14. (a) Show that $\log(\sqrt{10}\,P)$ is half way between $\log P$ and $\log(10\,P)$.

(b) Find $\sqrt{10}$.

(c) Use your answers to (a) and (b) to explain the statement: *Multiplication by 3 corresponds roughly to half a step on a logarithmic scale.*

15. In *Disturbing the Universe* Freeman Dyson writes:

> In 1961 and 1962 the United States and the Soviet Union exploded more bombs than ever before. Many of them were megaton bombs, and the fallout radioactivity was rising alarmingly all over the Northern Hemisphere. One quiet evening in my office at ACDA, I collected information about the tests and drew a simple diagram to show what was happening. From left to right I plotted the years from 1945 to 1962. Above each year I plotted vertically the cumulative total number of all bombs exploded from 1945 to that year. As soon as the diagram was finished, the situation became clear. The curve of cumulative bomb totals was an almost exact exponential, all the way from 1945 to 1962, with a doubling time of three years. A simple explanation suggested itself for this doubling every three years. It takes roughly three years to plan and carry out a bomb test. Suppose that every completed bomb test raises two new questions which have to be answered by two new bomb tests three years later. Then the exponential curve is explained. Having discovered this profound truth about bomb tests, I was ready to draw the consequences. Some questions have to remain unanswered. At some point, we have to stop. That evening I accepted for the first time the inevitability of a test ban.

The first test was at Alamagordo, New Mexico, in 1945. Had there been no test-ban treaty, how many tests would have taken place this year? When was the test-ban treaty signed?

CHAPTER 13

The Natural Base e

> There are more things in heaven and earth, Horatio,
> Than are dreamt of in your philosophy.
>
> – *Hamlet*, Act I, Scene V

Interest compounded periodically and continuously
A definition of *e* / Formulas for exponential growth, half-lives, and doubling times rewritten using *e*

EXAMPLE 1. COMPOUND INTEREST

Neighborhood Bank pays 5.25 percent annual interest on money in savings accounts, and compounds interest monthly. If you invest D dollars now, what will your bank balance be in 6 months? In a year?

To answer these questions we must understand what the word *compounds* means in the problem. Think about the difference between an annual inflation rate of 5.25% and simple interest at an annual rate of 5.25%. Each calls for the use of an annual growth factor of 1.0525, but inflation is going on continuously, while the bank pays interest only at the end of the deposit year. After a half year an item which used to cost C dollars cost $C * 1.0525^{0.5}$, but D dollars invested at the bank would still be only D dollars. The bank balance jumps suddenly to $D * 1.0525$ at the end of the year, but does not change until then, unless the bank *compounds your interest*. Neighborhood Bank compounds monthly. That means that each month they add to your account the interest earned during that month. After one month you are entitled to $^1/_{12}$ of the year's interest, or $(5.25/12)\%$ of the amount you have in the bank. That 0.4375% corresponds to a *monthly* growth factor of

$$1 + \frac{0.0525}{12} = 1.004375.$$

Thus in 6 months D dollars will be worth

$$D * 1.004375^6 = D * 1.0265.$$

In M months, D dollars will be worth

$$D * 1.004375^M. \tag{13.1}$$

To find the balance in the account after one year, substitute $M = 12$ in Equation 13.1:

$$\begin{aligned} \text{1-year balance} &= D * 1.004375^{12} \\ &= D * 1.0538. \end{aligned} \tag{13.2}$$

The growth factor in Equation 13.2 is 1.0538, which corresponds to an annual interest rate of 5.38%. Thus the monthly compounding, which pays interest on accumulating interest as well as on the principal, makes the *nominal* rate of 5.25% offered by the bank an *effective* rate of 5.38%. *Nominal* means *in name only*. The nominal rate is what the bank says it offers, before it considers the effects of compounding.

You really get the effective rate, which is higher because the interest is compounded. The second column of Table 13.1 summarizes what we've learned about 5.25% nominal annual interest, compounded monthly. The third column in Table 13.1 gives formulas with which we can make similar calculations for other nominal rates and different compounding frequencies. Suppose the nominal interest rate is r; we have just seen what happens when $r = 0.0525$, or 5.25%. Then the nominal annual growth factor is $1 + r$. This can be effectively increased by compounding. Suppose that compounding occurs N times per year; we have just seen what happens when $N = 12$. $N = 52$ would indicate weekly

Table 13.1 Compound interest

Nominal annual rate	5.25%	r
Nominal annual growth factor	1.0525	$1 + r$
Number of compoundings	12	N
Effective annual growth factor, G	1.00538	$\left(1 + \frac{r}{N}\right)^N$
Effective annual rate	5.38%	effective g.f. −1

compounding. In each case the interest rate for a single period, be it a month or a week or $1/N$ of a year, is r/N. The growth factor for that period is $1 + \frac{r}{N}$. Since there are N compoundings in a year, that growth factor works N times to produce the effective annual growth factor G:

$$G = \left(1 + \frac{r}{N}\right)^N.$$

For example, when the nominal annual interest rate is 11% and interest is compounded weekly, the effective annual growth factor is given by

$$G = \left(1 + \frac{0.11}{52}\right)^{52} = 1.1161.$$

Thus the effective annual rate is 11.61%.

EXAMPLE 2. NEIGHBORHOOD BANK VS. AGGRESSIVE BANK

Federal law limits the nominal annual interest rate payable on Special Certificates to 8%. Neighborhood and Aggressive Banks each want you to buy their certificates. They encourage you by compounding their 8% interest often, so that they can offer higher effective rates without violating the law. Who wins the competition?

Suppose Neighborhood begins by offering to compound quarterly. The effective annual growth factor is then

$$\left(1 + \frac{0.08}{4}\right)^4 = 1.0824,$$

which corresponds to an effective rate of 8.24%. Aggressive Bank retaliates by compounding monthly, so that they can promise an effective annual growth factor of

$$\left(1 + \frac{0.08}{12}\right)^{12} = 1.0830,$$

and thus an effective yield of 8.30%.

The folks at Neighborhood can see the pattern: Aggressive will top each of their offers. If they advertise weekly compounding, Aggressive will compound daily. If they try hourly, Aggressive will compound every minute. Neighborhood has hired us to study whether they can make a *best possible offer*.

Table 13.2 Effect of compounding on 8% nominal annual interest

Number of compoundings, N	Effective annual growth factor, G	Rate (%)
1 (end of year)	1.080 000	8.0000
4 (quarterly)	1.082 432	8.2432
12 (monthly)	1.083 000	8.3000
365 (daily)	1.083 278	8.3278
8760 (hourly)	1.083 283	8.3283

We begin by preparing Table 13.2, which shows the effective annual growth factor and rate when 8% nominal annual interest is compounded N times each year. As we suspect, the more often the bank compounds (the larger N becomes), the larger is the effective annual growth factor G. But the speed at which G increases with increasing N seems to slow down. When N increases from 4 to 12, G increases from 1.082 432 to 1.083 000, a change of $5.68 * 10^{-4}$. But when N increases from 365 to 8760, G changes by only $5 * 10^{-6}$. The value of G seems to level off. Let us calculate G for some enormous N, in order to test this hunch about leveling off. Suppose $N = 10^9$, which corresponds to compounding approximately every 30 milliseconds. Then:

$$\begin{aligned} G &= \left(1 + \frac{0.08}{10^9}\right)^{10^9} \\ &= (1 + 0.000\,000\,000\,08)^{10^9}, \end{aligned} \tag{13.3}$$

which our calculator tells us is

$$(1.000\,000\,00)^{10^9} = 1.000\,000\,00. \tag{13.4}$$

Something is wrong! This means no growth at all. We expected an answer just a little bigger than 1.083 283.

> The problem is not in our understanding. Equation 13.3 is perfectly correct. But we have exceeded the capabilities of the calculator. It can handle $0.08/(10^9)$ by switching to scientific notation and displaying $8 * 10^{-11}$. But when we add 1 to this small number the machine has no choice but to round the answer to about eight decimal places. (Exactly how many depends on the particular kind of calculator you have.) That rounding error produces the wrong answer in Equation 13.4.

But in spite of our calculator's inability to find G when compounding takes place every 30 milliseconds, we still believe that G is leveling off.

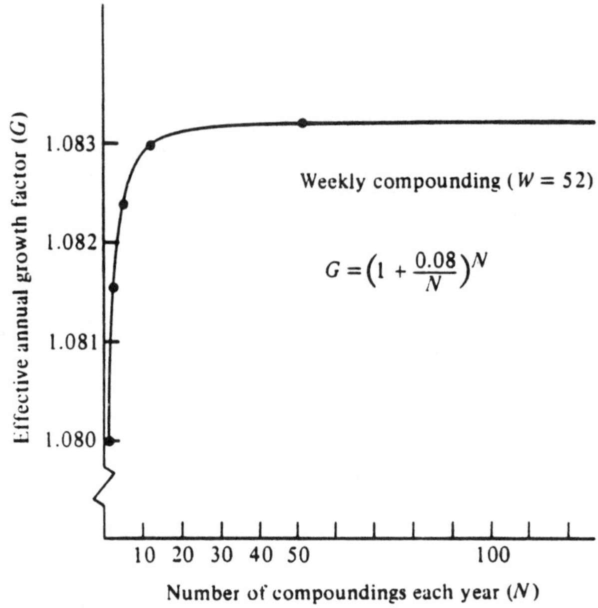

Figure 13.1 The effect of compounding interest

The graph in Figure 13.1 shows how G depends on N. There the leveling off is strikingly visible.

Neighborhood Bank should find the value of G at which the value of

$$\left(1 + \frac{0.08}{N}\right)^N \tag{13.5}$$

levels off as N gets bigger and bigger. They can then promise their customers the effective rate corresponding to that annual growth factor, which will be a shade more than 1.083 283. They will defeat Aggressive, and still satisfy federal law by telling the banking commission that they are *compounding continuously*.

Just one problem remains. How do they know exactly what the limit is in expression (13.5), when calculators fail if N is very large? This difficult question is the subject of the next example.

EXAMPLE 3. THE NATURAL BASE *e*

Suppose a bank offers a nominal annual interest rate of r, and compounds continuously. How can we compute the effective annual growth factor G?

Table 13.3 How the effective growth factor in nearly continuous compounding depends on the nominal rate

Nominal rate, r	$\left(1 + \frac{r}{3000}\right)^{3000}$ (which is nearly G)
0.05	1.051
0.08	1.083
0.1	1.105
0.15	1.162
0.2	1.221
0.5	1.649
0.8	2.225
1.0	2.718

We know that G is the number at which the expression

$$\left(1 + \frac{r}{N}\right)^{N} \tag{13.6}$$

levels off as N grows indefinitely. If we take a particular large value of N, then expression (13.6) tells us an approximate value for G. Table 13.3 shows such approximations for several values of r, using $N = 3000$. This corresponds to compounding about once every three hours. To improve the approximations we must take larger and larger values of N. However, we know from our experience with $N = 10^9$ in Example 2 that our calculators will run into trouble. We need an entirely different strategy.

In the last line in Table 13.3 $r = 1$. This represents a nominal annual interest rate of 100%, or a nominal annual growth factor of 2. Now no bank offers to double your money in a year. Nevertheless, we will find it useful to speculate about what would happen if one did—and then compounded interest as well. The result would be even better than doubling. We see from Table 13.3 that the effective annual growth factor would be approximately 2.718. The exact answer has an official name, e.

The number e is the number at which

$$\left(1 + \frac{1}{N}\right)^{N} \tag{13.7}$$

levels off as N grows indefinitely.

The value of e, correct to 13 decimal places, is

$$e = 2.718\ 281\ 828\ 459\ 0. \tag{13.8}$$

The number e, like π, has a decimal expansion that neither comes out even nor repeats. (The sequence, . . . 18281828 . . . , which occurs in places 2 through 9, looks like a repeating block, but this is a coincidence.)

> The value of e in Equation 13.8 was not found by looking at $(1 + (1/N))^N$ for large values of N. That method would work in principle, but fails in practice because of rounding errors. There are other methods, and your calculator knows one. If it has an $[e^x]$ key, pressing 1 $[e^x]$ will display the value of e. If no $[e^x]$ key is found, try
>
> 1 [INV] [LN] or 1 [2ND] [LN].
>
> Before proceeding, be sure you know how to make your calculator tell you the value of e.
>
> Also be sure you understand the difference between the *meaning* of e, which is given by definition 13.7, and its *value*, which is given (approximately) by Equation 13.8. That is like the distinction between the *meaning* of log 2, which is the solution to the equation $10^? = 2$, and its *value*, which is about 0.301 03.

We shall find many uses for e in the following chapters. We can do more with it than find the return when money is invested at 100%, continuously compounded. We start by showing how we can use it to calculate rapidly the exact values for column 2 of Table 13.3:

> **Continuously compounding interest at the nominal annual rate r produces an effective annual growth factor of e^r.**

We shall soon see why this is true. First, however, use it to check the second column of Table 13.3 by calculating $e^{0.05}$, $e^{0.08}$, and so on. In each case the answer, which is the exact effective growth factor, should be slightly larger than the approximation in the table.

> The key you used earlier to find e really finds e^x. Now use the $[e^x]$ key (or one of the pairs [INV] [LN] or [2ND] [LN]) to check that
>
> $$e^{0.05} = 1.051\,271.$$
>
> Never use the $[y^x]$ key to calculate the value of e^x.

We still have one task left in this example. We have claimed that the effective annual growth factor for Neighborhood Bank's continuously compounded nominal annual 8% interest rate is $e^{0.08} = 1.083\,287\,1$, which seems to agree with the evidence we collected in Table 13.2. But why should this be true? What makes the expression

$$\left(1 + \frac{0.08}{N}\right)^N \tag{13.9}$$

level off at $e^{0.08}$ when N grows very large?

The argument that follows is probably the subtlest in this book. Start with some really large number M; you might want to think of M as about 10^{24}. Then let $N = 0.08M$. N is only 8% of M, but it's pretty big too, and so if we substitute $N = 0.08M$ in expression (13.9) we will have a number near the leveling off value G.

$$\begin{aligned} G &\sim \left(1 + \frac{0.08}{0.08\,M}\right)^{0.08M} \\ &= \left(1 + \frac{1}{M}\right)^{M * 0.08} \\ &= \left[\left(1 + \frac{1}{M}\right)^M\right]^{0.08} \\ &\sim e^{0.08}. \end{aligned} \tag{13.10}$$

You should be able to supply the reasons for all but the last crucial step. This one is true because M is large, so that $(1 + 1/M)^M$ is very nearly e. The larger M is, the closer each of the two approximations in Equation 13.10 is to an equality. Thus the final answer to Neighborhood Bank's question is: By compounding the 8% nominal annual interest continuously, they can achieve an effective annual growth factor of

$$e^{0.08} = 1.083\,287\,1,$$

which corresponds to an effective annual interest rate of 8.328 71%. Table 13.2 shows that this is a little better than what they could offer with hourly compounding.

We have just found the ultimate growth factor G when $r = 0.08$. Next we generalize the argument in Equation 13.10, so that we can find G in terms of r when the numerical value of r is not known. Start with a very large number M: so large that $N = rM$ is still huge.

Then

$$
\begin{aligned}
G &\sim \left(1 + \frac{r}{\text{huge}}\right)^{\text{huge}} \\
&= \left(1 + \frac{r}{rM}\right)^{rM} \\
&= \left(1 + \frac{1}{M}\right)^{Mr} \\
&= \left[\left(1 + \frac{1}{M}\right)^{M}\right]^{r} \\
&\sim e^{r}
\end{aligned}
$$

EXAMPLE 4. NATURAL LOGARITHMS

What nominal annual interest rate will, when continuously compounded, give an effective annual growth factor of 1.17?

The answer is certainly less than 17%, because that nominal rate would have a growth factor of 1.17 with no compounding at all. To find the rate r we must solve the equation

$$e^{r} = 1.17 \tag{13.11}$$

for r. Use logarithms to get:

$$r * \log e = \log 1.17$$

and so

$$
\begin{aligned}
r &= \frac{\log 1.17}{\log e} \\
&= \frac{\log 1.17}{\log 2.718} \\
&= 0.157.
\end{aligned}
$$

Thus an effective growth factor of 1.17 will be produced by a nominal annual rate of 15.7%.

But we can solve Equation 13.11 for r in an easier way. There is a logarithm that undoes e^{x} just as log undoes 10^{x}. That logarithm is called the *natural logarithm*, abbreviated *ln*. (That's LN, not IN.) That is to say:

$$\ln Q \quad \text{is the name we give to the solution to} \quad e^{?} = Q. \qquad (13.12)$$

(Look back at Equation 11.5.) In other words, the equations

$$Q = e^T \qquad (13.13)$$

and

$$T = \ln Q \qquad (13.14)$$

say exactly the same thing. (Compare Equations 11.6 and 11.7.) In this example we can solve Equation 13.11 by calculating

$$r = \ln(1.17) = 0.157.$$

> Your calculator finds natural logarithms just as easily as common ones. Simply use the [LN] rather than the [LOG] key. On many calculators the fact that the operations "*e* to the power ___" and "natural logarithm of ___" invert one another is reinforced by having the same key do both jobs. One of them is signaled by first using the [INV] or [2ND] key.

All the rules for common logarithms have their analogues for natural logarithms. They are summarized in Table 13.4.

Table 13.4 Properties of logarithms

$10^{\log Q} = Q$		$e^{\ln Q} = Q$
$\log(10^T) = T$		$\ln(e^T) = T$
$\log(10) = 1$		$\ln(e) = 1$
$10^0 = 1$		$e^0 = 1$
$\log 1 = 0$		$\ln 1 = 0$
	If $T > 1$ then	
$\log T > 0$		$\ln T > 0$
	If $T < 1$ then	
$\log T < 0$		$\ln T < 0$
$\log(A * B) = \log A + \log B$		$\ln(A * B) = \ln A + \ln B$
$\log(K^T) = T \log K$		$\ln(K^T) = T \ln K$
$\log\left(\frac{1}{Q}\right) = -\log Q$		$\ln\left(\frac{1}{Q}\right) = -\ln Q$

EXAMPLE 5. RESHAPING THE EQUATION FOR EXPONENTIAL GROWTH

Our friends at Neighborhood Bank have taken our advice. They offer savings certificates at a nominal annual interest rate of 8 percent, and compound the interest continuously. If we invest now, what will our balance be in t years?

The balance B depends on the time t. The relationship is exponential:

$$B = B(0) * K^t, \tag{13.15}$$

where $B(0)$ is the initial investment, and K is the effective annual growth factor. We now know that

$$K = e^{0.08} = 1.0833,$$

and so Equation 13.15 becomes

$$B = B(0) * 1.0833^t.$$

But there is another way to say this which leaves e in the equation. Substituting $K = e^{0.08}$ in Equation 13.15 gives

$$\begin{aligned} B &= B(0) * (e^{0.08})^t \\ &= B(0) * e^{0.08t}. \end{aligned} \tag{13.16}$$

In general, when the nominal annual interest rate is r rather than 0.08, substitute $K = e^r$ in Equation 13.15. Then

$$B = B(0) * e^{rt}. \tag{13.17}$$

It is a measure of the importance of the number e that throughout the literature of science, growth factors K are always written in the form e^r, even when the problem has nothing to do with compound interest. The form of Equation 13.17 is preferred to that of Equation 13.15. For example, in Chapter 9 we discovered that the equation governing the weight W of a particular colony of bacteria was

$$W = 3.8 * 1.7^T. \tag{13.18}$$

To rewrite this equation so that the growth factor is a power of e we must solve $e^r = 1.7$ for r. Fortunately, we know how:

$$r = \ln(1.7) = 0.53,$$

and so Equation 13.18 becomes

$$W = 3.8 * e^{0.53T}. \tag{13.19}$$

The principle is: To rewrite

$$P = P(0) * K^t \tag{13.20}$$

in the form

$$P = P(0) * e^{rt} \tag{13.21}$$

we let

$$r = \ln K. \tag{13.22}$$

If we are given Equation 13.21, and wish to recover the growth factor K in Equation 13.20, we just undo the calculation in Equation 13.22:

$$K = e^r.$$

For example, to rewrite

$$W = 1.235 * 0.9174^D,$$

(Equation 9.16, describing the decay of iodine-131), first calculate

$$\ln(0.9174) = -0.086\,21.$$

Then write

$$W = 1.235 * e^{-0.086\,21\,D}. \tag{13.23}$$

Notice that the rate constant −0.086 21 in the exponent is negative, because the growth factor 0.9174 is less than 1. The amount of iodine-131 decreases as the number of days D increases.

EXAMPLE 6. DOUBLING TIMES AND HALF-LIVES USING e

What is the doubling time for the population of bacteria whose weight W in milligrams grows exponentially following the rule

$$W = W(0) * e^{0.53T}? \tag{13.24}$$

One way to answer this question is to calculate the growth factor $K = e^{0.53}$ and then use the method we learned in Chapter 12:

$$\text{doubling time} = \frac{\log 2}{\log K}.$$

But when the growth equation is written in terms of e, there is a shorter way. We are looking for the time that gives a growth factor of 2. Equation 13.24 tells us that this is the value of T for which

$$e^{0.53T} = 2. \tag{13.25}$$

To solve this for T, rewrite it using natural logarithms instead of powers of e:

$$0.53T = \ln 2 = 0.6931$$

so

$$T = \text{doubling time} = \frac{0.6931}{0.53} = 1.3 \text{ days}.$$

The general formula this exemplifies tells us how to calculate the doubling time for the exponentially growing quantity Q with rate constant r. If

$$Q(t) = Q(0) * e^{rt},$$

then

$$\begin{aligned} \text{doubling time} &= \frac{\ln 2}{r} \\ &= \frac{0.6931}{r} \\ &\sim \frac{0.7}{r}. \end{aligned} \tag{13.26}$$

That's nicer than the corresponding formula:

$$\text{doubling time} \sim \frac{0.3}{\log K} \tag{13.27}$$

because it has no logarithms in it. Equation 13.26 tells you immediately that 7% interest doubles your money in 10 years. It's harder to extract the same information from Equation 13.27.

To see how to calculate half-lives using e, look at the decay of iodine-131, which is governed by Equation 13.23. We seek the value of D for which

$$e^{-0.086\,21D} = \frac{1}{2}.$$

This says that

$$-0.086\,21\,D = \ln\left(\frac{1}{2}\right) = -\ln(2),$$

so

$$D = \frac{0.6931}{0.086\,21} = 8.04 \text{ days}.$$

The general formula for the half-life of a substance decaying according to the rule

$$Q(t) = Q(0) * e^{rt}$$

is

$$\begin{aligned} \text{half-life} &= \frac{-\ln 2}{r} \\ &\sim \frac{-0.7}{r}. \end{aligned} \tag{13.28}$$

The answer in Equation 13.28 appears to be negative, but is actually positive, because, since Q is decreasing, r is negative.

PROBLEMS

1. Compute the effective annual growth factor and the effective annual interest rate if the nominal rate is 6 percent and compounding occurs

(a) annually; (b) semiannually; (c) quarterly;
(d) monthly; (e) daily; (f) hourly;
(g) every minute.

Look for and identify the point at which rounding error in your calculator produces an answer you can be sure is wrong.

2. Answer the previous question when the nominal rate is 10% ($r = 0.1$); and when it is 100% ($r = 1$).

*3. Find the doubling time for money invested at 9.47% nominal annual interest compounded monthly.

4. Try using the [y^x] key to find [e^x] for $x = 1.22$. Why is this a bad idea?

5. Find the effective annual growth factor and the effective annual rate for continuously compounded interest at nominal rates of 2%, 3%, 5%, 7%, 10%, 12%, and 18%.

6. (a) Find the doubling time for continuously compounded interest at each of the nominal rates in the previous problem.
   (b) Draw a graph showing how doubling time depends on nominal rate.

7. Bankers use the *rule of 70* to calculate doubling times: Money growing at a rate of r% per year doubles in $70/r$ years. Use the rule of 70 to answer Problem 6(a). Why does this rule work?

*8. Which is better for you: 13% nominal annual interest compounded monthly, or 12.9% nominal annual interest compounded continuously? Which is better for the bank?

9. Banks use the formula

$$P = P(0) * \left(1 + \frac{r}{360}\right)^D$$

to compute what a principal of $P(0)$ dollars is worth after D days in what they call *daily compounding*. Compute the true (that is, *effective*) annual rate for a nominal rate of $r = 6\% = 0.06$ according to:

(a) the bank's formula;
(b) continuous compounding;
(c) real daily compounding (365 days/year).
(d) Rank these rates in order. Which is best for you?

10. Here are two investment plans:

River Bank	13.65% compounded quarterly
Faithless Trust	13.51% compounded continuously

(a) Which plan gives a better return? Explain.

(b) Bartholomew D. C. Manypenny invests \$500 in each plan. In each case express the value of his investment after Y years in the form $500e^{kY}$.

(c) How long will a \$500 investment take to grow to \$650 using the River Bank plan? How long at Faithless Trust?

11. A merchant put \$50 aside for each of his newborn twin sons. One deposit went to a bank in Syracuse, the other to a bank in Ephesus. Each bank compounded the interest monthly. Now the boys have come of age, at 21, and it turns out that they both have the same bank balance. The bank in Syracuse paid 5% interest (nominal annual rate) for the first 8 years and then 8% interest for the next 13 years. The bank at Ephesus paid interest at the same nominal rate the whole time.

(a) What is each bank balance now?

(b) What was the rate at Ephesus?

*12. A five-year account at a savings bank pays 8% annual interest. How much should be deposited now so that the balance in 20 years will be \$50 000 if

(a) interest is compounded weekly?

(b) interest is compounded continuously?

*13. A Money Market certificate with nominal annual rate of 9.495% promises an effective yield of 9.860%.

(a) How long will it take to double your money?

(b) How often, approximately, is interest compounded?

14. The school budget in a suburban town was \$600 000 in 1960 and \$2 million in 1980.

(a) Suppose the growth was exponential. Write a formula, using e, showing how the school budget changed with time. Draw the graph. Predict the 1990 budget. When will the budget reach \$2.5 million?

(b) Answer the questions in (a) if the growth was linear. Draw the graph on the same set of axes.

(c) Suppose inflation has averaged 7% per year since 1960. Has the school budget increased faster than inflation? Could the town buy more for \$2 million in 1980 than it could for \$600 000 in 1960?

15. An accountant can choose either of the following depreciation plans for a \$25 000 truck. The age t is measured in years, the value V in thousands of dollars.

$$V = 25 - 1.6t \text{ (linear)}$$

or

$$V = 25e^{-0.11t} \text{ (exponential)}$$

(a) Graph both depreciation curves on the same set of axes. For each plan, answer the following questions.
(1) What is the value of the truck in 5, 10, 15, and 25 years?
(2) When is the truck worth half its original value?
(3) When is it worth \$15 000?
(4) When is it worthless?

(b) At what ages do the plans give the same value V?

*16. An HP calculator cost \$98.50 in 1976 and \$73.00 three years later. Assuming that the price drops at an exponential rate, write an equation using e that shows how the cost changes with time. What does the calculator cost now?

17. Rewrite each of the following exponential equations using e:
(a) $W = 8.9 * 1.71^t$
(b) $N = 2870 * 0.948^t$
(c) $R = R(0) * 2.7183^D$
(d) $P = P(0) * 1.003^Y$
(This problem anticipates Example 1 of Chapter 14.)

*18. Rewrite each of the following exponential equations without using e, so that the growth factor is explicitly visible:
(a) $S = 1.38 * e^{1.4t}$
(b) $T = 0.0063 * e^{-0.31y}$
(c) $M = M(0) * e^t$
(d) $Z = Z(0) * e^{0.004d}$
(This problem anticipates Example 1 of Chapter 14.)

19. Rewrite the growth and decay formulas from Example 1 of Chapter 9, and Examples 1 and 2 of Chapter 10, using e.

20. Redo each of the problems in Chapter 10 that calls for a formula describing exponential growth or decay. This time use e.

21. The population of the world is about 4 billion people. It has been growing at about 2% per year. (See Problem 11 of Chapter 9.)

    (a) Find an exponential formula using e for the population.

    (b) How many people will there be in the year 2000? In the year 3000?

    (c) How many were there in 1900?

    (d) How many were there 600 years ago? (See Problem 15 of Chapter 9.)

    (e) How long does it take the world's population to double?

22. The short-lived isotope helium-5 (^{5}He) has a half-life of $2.4 * 10^{-21}$ seconds.

    (a) Express this in microseconds and in nanoseconds.

    (b) Write a formula using e that shows how the amount of ^{5}He changes with time.

    (c) How long will it take for a quantity of ^{5}He to decay to one thousandth of its original amount? To one millionth?

23. Explain why, if a quantity Q grows exponentially with time, then its natural logarithm $\ln(Q)$ grows linearly with time.

24. (a) Complete the following table.

r	$\log r$	$\ln r$
0.1		
$1/e$		
0.5		
1		
2		
e		
5		
10		

    (b) Draw a graph that shows how $\log r$ depends on $\ln r$. What conclusion do you draw?

    (c) Show algebraically that $\log(r)$ is proportional to $\ln(r)$, and that the proportionality constant is $\log(e) \sim 0.43$. (*Hint*: Start with the equation: $r = e^{(\ln r)}$ and take logarithms.)

25. Solve for L: $e^L = 10^{100}$
What is the largest number L for which your calculator can compute e^L?

26. Find t if: $28.7 = 16.3\, e^{1.81t}$

*27. Solve for t in terms of P: $P = P(0)\, e^{kt}$

*28. Find H if: $e^{(1+\ln H)} = 2$

CHAPTER 14

Growth

> Put not yourself into amazement how these things should be;
> all difficulties are but easy when they are known.
>
> – *Measure for Measure*, Act IV, Scene 3

Why e is natural / The slope of exponential curves
Other growth models

EXAMPLE 1. e^x FOR SMALL VALUES OF x

Observe the values of e^x when x is close to 0, and explain the observations.

We shall see in Example 2 how this numerical exercise, which is not like most of the other examples we have studied, helps us understand exponential growth better.

We begin by constructing Table 14.1, which gives the value of e^x for several small values of x. The conclusion is inescapable. When x is small,

$$e^x \sim 1 + x. \tag{14.1}$$

We can say even more. First, the table shows that the approximation is good when x is negative as well as positive. Second, the approximation is a little too small–e^x is close to but just a bit larger than $1 + x$. Third, the smaller the value of x, the better the approximation. When x is between -0.003 and 0.003 the approximation is good to 5 decimal places; when $x = 0$ it is exact.

Figure 14.1 explains the approximations in Table 14.1. It shows the graph of the curve

$$Q = e^x$$

Table 14.1 e^x when x is small

x	e^x	$1+x$
0.2	1.221 40	1.200 00
0.1	1.105 17	1.100 00
0.05	1.051 27	1.050 00
0.04	1.040 81	1.040 00
0.03	1.030 45	1.030 00
0.02	1.020 20	1.020 00
0.01	1.010 05	1.010 00
0.005	1.005 01	1.005 00
0.004	1.004 01	1.004 00
0.003	1.003 00	1.003 00
−0.2	0.818 73	0.800 00
−0.1	0.904 84	0.900 00
−0.05	0.951 23	0.950 00
−0.04	0.960 79	0.960 00
−0.03	0.970 45	0.970 00
−0.02	0.980 20	0.980 00
−0.01	0.990 05	0.990 00
−0.005	0.995 01	0.995 00
−0.004	0.996 01	0.996 00
−0.003	0.997 00	0.997 00

and of the line

$$Q = 1 + x.$$

The line, with intercept 1 and slope 1, is *tangent* to the curve when $x = 0$ and $Q = 1$. When x is close to 0 the curve is close to but just above its tangent.

EXAMPLE 2. THE RATE OF EXPONENTIAL GROWTH

In Example 2 of Chapter 9 we studied a colony of bacteria that grew exponentially according to the rule

$$W = 3.8e^{0.53t}, \qquad (14.2)$$

where W was measured in milligrams and t in hours. Now consider its growth rate in mg/hr.

We have never before encountered this kind of question about exponential growth. Since the units of the required rate are mg/hr we are looking for a slope. Until now we used slopes only when studying

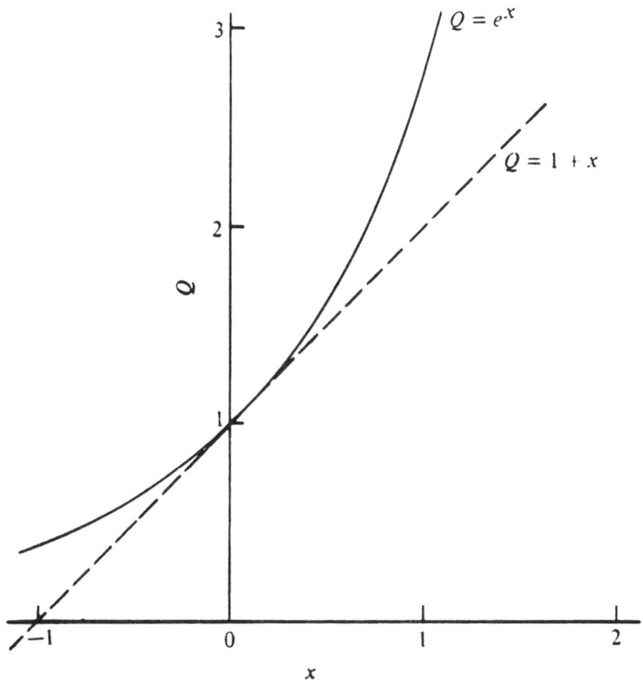

Figure 14.1 Establishing an approximation

linear models, because then the slope did not change with time. In exponential growth the slope keeps changing. Exponential curves, like the one in Figure 14.2, rise more and more steeply as time goes on. It is this changing slope that we wish to study, so we give it a name. Let R be the rate at which the population W grows. As t increases, W grows at a faster and faster rate; that is, R grows too. We shall discover that the rate R at which the population changes is *proportional* to the size of the population. The proportionality constant turns out to be the coefficient 0.53 in the exponent in Equation 14.2 Thus the slope R of the growth curve in Figure 14.2 is given by the equation

$$R = 0.53W. \tag{14.3}$$

For example, at time $t = 1.5$ hours, the colony weighs

$$W(1.5) = 3.8e^{0.53 * 1.5} = 8.4 \text{ mg}$$

and is growing at the rate

$$R = 0.53W = 0.53 * 8.4 = 4.5 \text{ mg/hr}.$$

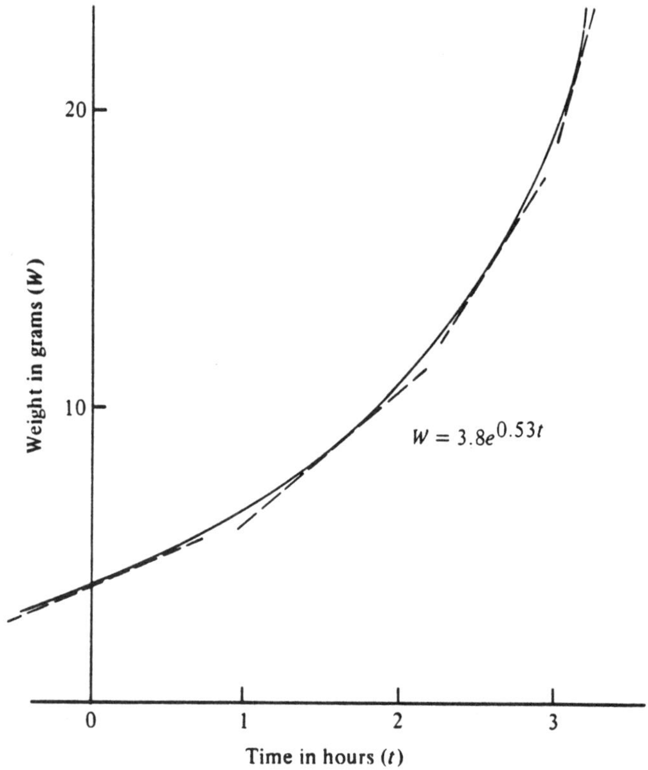

Figure 14.2 Tangents to a growth curve

At time $t = 2$ hours the weight is

$$W(2) = 10.7 \text{ mg}$$

and the growth rate is

$$R = 0.53W = 0.53 * 10.7 = 5.8 \text{ mg/hr.}$$

Let us see where Equation 14.3 comes from, first in the special case $t = 1.5$ hours. To find the slope R we must calculate the ratio

$$\frac{\text{change in } W}{\text{change in } t}. \tag{14.4}$$

If we use a very short time interval in Equation 14.4 the exponential growth will be nearly linear. Let us see what happens between times $t = 1.5$ and $t = 1.5 + h$ hours, where h is very small. Then

$$\text{change in } t = h. \tag{14.5}$$

It is a little harder to find the change in W. We first find its new value, $W(1.5 + h)$. Equation 14.2 implies that the h hour growth factor is $e^{0.53h}$. Thus at time $t = 1.5 + h$,

$$W(1.5 + h) = W(1.5) * e^{0.53h}. \tag{14.6}$$

Therefore,

$$\begin{aligned} \text{change in } W &= W(1.5 + h) - W(1.5) \\ &= W(1.5) * e^{0.53h} - W(1.5) \\ &= W(1.5) * (e^{0.53h} - 1). \end{aligned} \tag{14.7}$$

Remember that h is small. Thus we can use Equation 14.1 to make the replacement

$$e^{0.53h} \sim 1 + 0.53h$$

in Equation 14.7. This yields

$$\begin{aligned} \text{change in } W &\sim W(1.5) * (1 + 0.53h - 1) \\ &\sim W(1.5) * (0.53h). \end{aligned} \tag{14.8}$$

Finally, combining Equations 14.4, 14.5, and 14.8, we get

$$\begin{aligned} R &= \frac{\text{change in } W}{\text{change in } t} \\ &\sim \frac{W(1.5) * 0.53h}{h} \\ &= 0.53 * W(1.5), \end{aligned} \tag{14.9}$$

which confirms our belief in Equation 14.3. This elaborate algebra is independent of the fact that we have been concentrating on the particular time $t = 1.5$ hours. To deduce Equation 14.3 in general, just replace the number 1.5 by the variable t in Equations 14.6 through 14.9.

We have actually shown still more. In our argument we made no use of the particular numbers 3.8 and 0.53. They could just as easily have been written as constants $W(0)$ and r. Thus, whenever a quantity W grows exponentially, so that

$$W = W(0)e^{rt}, \tag{14.10}$$

it follows that W changes at a rate proportional to its value:

$$R = \text{rate of change of } W = rW. \tag{14.11}$$

Since in this example the units of R are mg/hr, and the units of W are mg, the units of the proportionality constant r are

$$\frac{\text{mg/hr}}{\text{mg}}, \quad \text{or} \quad \frac{1}{\text{hr}}, \quad \text{or} \quad \text{hr}^{-1}.$$

The relationship between Equations 14.10 and 14.11 is the hallmark of exponential growth. Either equation implies the other. When we write the exponential growth equation using e rather than an explicit growth factor, the crucial constant r appears in two significant places. In our argument we needed the approximation

$$e^x \sim 1 + x, \tag{14.12}$$

which we found in Example 1. It is that approximation, which is correct only for the number e, which makes e the natural number to use when discussing exponential growth.

EXAMPLE 3. WHERE THE WOLF HOWLS

Suppose that 60% of a population of Alaskan wolves is of breeding age, and that the breeding wolves produce an average of 0.8 cub per couple per year. Suppose further that the death rate for the population as a whole is 10% per year. How does the wolf population grow? What is its doubling time?

Let W represent the wolf population, which depends on the time t, measured in years. To answer the question we must understand what the various given rates mean. A population of size W will contain $0.6W$ wolves of breeding age, and hence $0.6W/2 = 0.3W$ couples. These couples produce

$$0.8\,\frac{\dfrac{\text{cub}}{\cancel{\text{couple}}}}{\text{year}} * 0.3W\,\cancel{\text{couple}} = 0.24\,\frac{\text{cub}}{\text{year}}. \tag{14.13}$$

Since 10% of the wolves die each year, the number of deaths each year is $0.1W$, and the death rate is

$$0.1W \text{ wolf/year.} \tag{14.14}$$

Let R stand for the rate of change of the wolf population W measured in wolves/year. Then, combining Equations 14.13 and 14.14, we get

$$R = 0.24W - 0.1W = 0.14W.$$

The rate of change of W is proportional to W. But that is just the situation we studied in Example 2. Hence the equation

$$W = W(0)e^{0.14t}$$

describes how W grows. The doubling time is

$$\frac{0.7}{0.14} = 5 \text{ years.} \tag{14.15}$$

Notice how nicely the units behave in Equation 14.15. The 0.7 is just ln(2); it has no units. The units of the proportionality constant 0.14 are 1/years.

EXAMPLE 4. FISH IN LAKE SKATUTIKEE

Lake Skatutikee can support a population of at most 1700 bass. It was stocked with 170 in 1950. Discuss how the population grew.

Let B stand for the bass population and t for the time in years since 1950. We shall study three possible models for the growth of B. Each is characterized by an assumption about the behavior of the rate R at which the population B changes.

Suppose that R is proportional to B, so that the larger B is the faster it grows. That is, suppose that

$$R = rB.$$

The work we did in Example 2 then tells us that B grows exponentially, so that

$$B = 170\, e^{rt}. \tag{14.16}$$

A reasonable value for the constant r is 0.25. This corresponds to 0.25

new fish per year per fish present, or one new fish each year for each four old fish. Then

$$B = 170\, e^{0.25t}. \tag{14.17}$$

Figure 14.3 shows this exponential growth curve exceeding the 1700-fish maximum in about 9 years.

But this is not a realistic growth model. We would expect, as the population B approaches the maximum, that food will become scarce and predators will multiply, causing the rate of growth to decrease. This expectation is met by a model in which the growth rate R is proportional to the *difference* between 1700 and B:

$$R = 0.25(1700 - B). \tag{14.18}$$

More complicated mathematics than we can now handle implies that B grows according to the rule

$$B = 1700(1 - 0.9e^{-0.25t}). \tag{14.19}$$

Even though we cannot present here the argument from calculus that shows how Equation 14.19 follows from Equation 14.18, we can

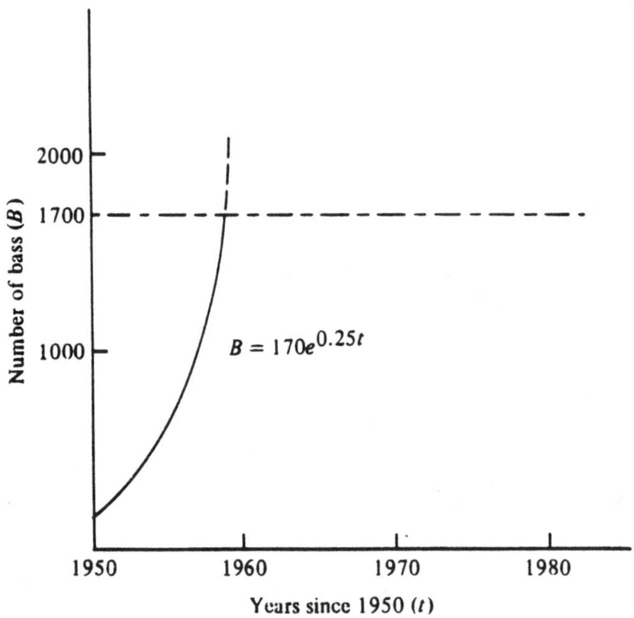

Figure 14.3 Bass population in Lake Skatutikee (I)

Table 14.2 Fish in Lake Skatutikee

t Years since 1950	B: bass population Model 1	Model 2
0	170	170
2	772	263
4	1137	394
6	1359	565
8	1493	766
10	1574	977
15	1664	1403
20	1690	1602
25	1697	1670
30	1699	1691

explore the consequences of Equation 14.19. The middle column in Table 14.2 lists some of the results; the solid curve in Figure 14.4 is the corresponding graph. These show B rising toward 1700, but more and more slowly as the limit is approached.

The model leading to Equation 14.19 has shortcomings too. When the population is small, the limiting value of 1700 fish should play no role at all, and the growth should be exponential: a small initial rate

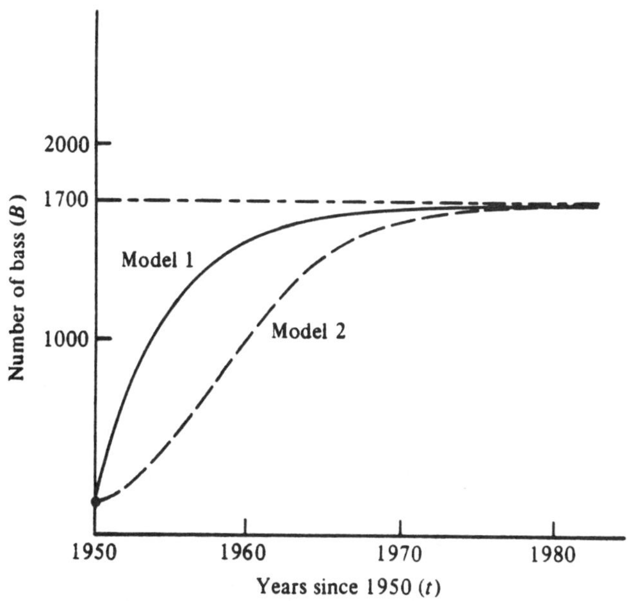

Figure 14.4 Bass population in Lake Skatutikee (II)

that increases with time. But the solid curve in Figure 14.4 is steepest when B is still small. To combine the models and construct a curve that looks exponential when t is small but levels off when t is large, we suppose the growth rate R is *jointly* proportional to B and to the difference $1700 - B$:

$$R = 0.25\,B(1700 - B). \tag{14.20}$$

Equation 14.20 gives a small growth rate R both for small populations, when B is small, and for populations near the limit, when $1700 - B$ is small.

It can be shown that Equation 14.20 implies that

$$B = 1700\,\frac{0.1\,e^{0.25t}}{1 + 0.1(e^{0.25t} - 1)}. \tag{14.21}$$

The third column in Table 14.2 shows the results of calculating B with Equation 14.21. The corresponding graph is the dotted line in Figure 14.4. It shows B rising slowly at first, then faster through the middle range, then more slowly as the limit is approached.

PROBLEMS

1. How close must x be to 0 to guarantee 4-decimal-place accuracy in the approximation

$$e^x \sim 1 + x?$$

2. Find numerical evidence for the approximations

$$2^x \sim 1 + (\ln 2)x$$

and

$$3^x \sim 1 + (\ln 3)x$$

when x is small. What pattern do these approximations suggest?

3. Find numerical evidence for an approximation to

$$\log(1 + x)$$

good for small values of x. Show how your approximation follows from Equation 14.1.

4. (a) Find numerical evidence for the approximation

$$\frac{1}{1-x} \sim 1 + x$$

when x is small.

(b) Use the algebraic fact that

$$(1-x)(1+x) = 1 - x^2$$

to explain this approximation.

5. (a) Find numerical evidence for the approximation

$$\sqrt{1+h} \sim 1 + \frac{h}{2}$$

when h is small.

(b) Use the algebraic fact that

$$\left(1 + \frac{h}{2}\right)^2 = 1 + h + \frac{1}{4}h^2$$

to explain this approximation.

6. When does B reach 1700 in the model given by Equation 14.14?

*7. The value V of a share of Wing Aero stock is appreciating according to the formula

$$V = 100\, e^{0.19Y},$$

where Y is measured in years since 1975, the year in which the stock was first sold.

(a) What did the stock first sell for?

(b) How much was it worth in 1982?

(c) How fast (in \$/year) was its value increasing when it was first issued? How fast in 1983?

(d) How fast will its value be increasing when it is worth \$1000? (You can answer this question without finding out when this occurs.)

8. At what rate (in number of people per year) was the world's population growing in 1900? What is the corresponding rate now? What will it be in the year 2000? (See Problem 21 of Chapter 13.)

*9. How does the rate of change R (in \$/year) of the value of the calculator we studied in Problem 16 of Chapter 13 depend on time? What was the rate R in 1976? What is it now?

10. If the rate R at which a quantity Q changes with time t is given by

$$R = 0.53Q,$$

and $Q(3.3) = 15.4$, find a formula showing explicitly how Q depends on t. Use this formula to calculate $Q(2.8)$.

*11. Bacteria reproduce asexually. Suppose that each bacterium produces k daughter cells per hour, and that the average individual's life span is h hours. Let $P(t)$ represent the population of a colony of bacteria at time t, measured in hours, and R the rate at which P is changing.

(a) Explain why there are about P/h bacteria deaths per hour.

(b) Write the equation that shows how R is proportional to P.

(c) Write the equation that shows how P depends on t.

(d) Express the doubling time of the population in terms of h and k.

12. Equation 14.19 is a special case of the growth model

$$B = M\,(1 - pe^{-rt}), \qquad (14.22)$$

where M is the maximum population the lake can support, $1 - p$ is the proportion of M giving the initial value $B(0)$, and r is a rate constant. In Example 3, $M = 1700$, $p = 0.9$ (since the lake was stocked with 170 fish, or 10% of its capacity), and $r = 0.25$.

(a) Solve Equation 14.19 for t in terms of B.

(b) Solve Equation 14.22 for t in terms of B, M, p, and r.

(c) Solve Equation 14.22 for each of the constants M, p, and r in turn.

(d) Suppose that $r = 0.25$ and $p = 0.9$. Investigate how the growth of B depends on M by drawing graphs for $M = 100$, 1500, 1700, 2000, and 2500–all on the same set of axes.

(e) Suppose that $M = 1700$ and $p = 0.9$. Investigate how the growth of B depends on r by drawing graphs for $r = 0.1, 0.2, 0.25, 0.3$, and 0.8–all on the same set of axes.

(f) Suppose that $M = 1700$ and $r = 0.25$. Investigate how the growth of B depends on p by drawing graphs for $p = 0.2, 0.7,$

0.9, 1.0, 1.1, and 1.5—all on the same set of axes. The last three curves may surprise you.

13. Equation 14.21 is a special case of the growth model

$$B = M \frac{pe^{rt}}{1 + p(e^{rt} - 1)}, \qquad (14.23)$$

where M is the maximum population the lake can support, p is the proportion of M giving the initial value $B(0)$, and r is a rate constant. In Example 3, $M = 1700$, $p = 0.1$ (since the lake was stocked with 170 fish, or 10% of its capacity), and $r = 0.25$.

(a) Solve Equation 14.21 for t in terms of B.

(b) Solve Equation 14.23 for t in terms of B, M, p, and r.

(c) Solve Equation 14.23 for each of the constants M, p, and r, in turn.

(d) Suppose that $r = 0.25$ and $p = 0.1$. Investigate how the growth of B depends on M by drawing graphs for $M = 100$, 1500, 1700, 2000, and 2500—all on the same set of axes.

(e) Suppose that $M = 1700$ and $p = 0.1$. Investigate how the growth of B depends on r by drawing graphs for $r = 0.1$, 0.2, 0.25, 0.3, and 0.8—all on the same set of axes.

(f) Suppose that $M = 1700$ and $r = 0.25$. Investigate how the growth of B depends on p by drawing graphs for $p = 0.2$, 0.7, 0.9, 1.0, 1.1, and 1.5—all on the same set of axes. The last three curves may surprise you.

14. Suppose that the temperature inside is I°F and the temperature outside is E°F. (E stands for environment. O is a poor choice for a variable name.) If you take your thermometer outside at time $t = 0$ min then at time t min it registers

$$T = E + (I - E)e^{-kt} \text{ °F}.$$

The constant k depends only on the particular thermometer you are using, not on the values of E and I.

(a) Suppose that when $I = 62$°F and $E = 30$°F you find that $T = 50$°F after $t = 2$ min. Find the constant k.

(b) Draw a graph showing how T depends on t.

(c) What is the temperature reading on the thermometer after 10 minutes?

(d) When does the temperature reading T reach 46°F?

(e) Suppose that $I = 70$°F and $E = 96$°F. When does the temperature reading reach 83°F? Do you notice a coincidence?

(f) Show algebraically that for any thermometer, the time it takes to reach the average temperature $(E + I)/2$ depends only on k and not on E or I. This time is called the half-life of the thermometer. George Mackey knows the half-life of his thermometer, so that he can find the outside temperature by reading his thermometer inside, taking it out for its half-life, and then doubling the recorded temperature change.

CHAPTER 15

Problems Using Exponential Techniques

As I was going to St. Ives
I met a man with seven wives.
Every wife had seven sacks,
Every sack had seven cats,
Every cat had seven kits.
Kits, cats, sacks, and wives,
How many were going to St. Ives?

– *Mother Goose*

1. Salt poisons bacteria.
   If 1.2 mg of salt is added to a colony of 10 000 bacteria, then in one hour the population grows to 11 500. If only 0.5 mg is added, then in one hour there are 12 500 bacteria.
   (a) In each case find a formula of the form

   $$P = P(0)e^{rt}.$$

   describing how the population P depends on time.
   (b) Suppose that the growth-rate constant r in (a) depends linearly on the quantity of salt added. Draw a graph exhibiting this dependence. Extrapolate to find how much salt should be added to produce a zero growth rate.
2. Depreciation vs. inflation.
   (a) A truck costs \$23 000 new and will be worth \$8000 when it is 8 years old. An accountant must decide whether to use a linear or an exponential model for the depreciation. Find formulas and graphs with which to compare the models. In particular, find out when each model gives the greater value, when the

value is \$15 000, when it is \$5000, and when the truck is worthless.

(b) Adjust your answers to (a) so as to take into account an inflationary increase of 9% per year in all dollar amounts. (*Hint*: Multiply each formula by an appropriate growth factor.)

(c) What inflation rate would cause the dollar value of the truck in the exponential model to remain constant?

3. Electric power consumption.
When time t is measured in years since 1920, P (the annual electric energy consumption in the United States) is measured in terawatt-hours per year (TWh) and W, the United States population, is measured in millions of people, then for the period 1920–1970 it was approximately true that:

$$P = 54\, e^{0.07t} \quad \text{and} \quad W = 106\, e^{0.013t}.$$

(a) Study these equations separately. Graph each, and graph the natural logarithms of each. (See Example 1 of Chapter 11.) Find the doubling times. Make predictions for this year and for the year 2000. Did the model accurately predict the 1980 population found by the census?

(b) The per capita consumption C is given by $C = P/W$. Thus in 1920, C was 0.51 TWh/million people.
   (1) How many KWh/person is this?
   (2) Find a formula for C of the form $C = C(0)e^{rt}$. (*Hint*: Use the laws of exponents in the expression P/W.)
   (3) Find C for 1930 and 1970, predictions for this year and the year 2000, both using the expression P/W directly and, as a check, using the formula found in (2).

(c) Suppose that Americans learned to conserve energy, so that starting in 1980 C remained constant. What would happen to P if W continued to grow as usual? Find a new formula for P and use it to revise your predictions for 2000. Draw a new curve for P, properly labeled, on the same graph as the old one.

(d) Suppose that the population ceased to grow after 1980 (that is, W remained constant) while C grew as usual. Answer the question asked in (c).

(e) Suppose that both C and W were constant after 1980. Answer the question asked in (c).

(f) Use your figures to determine the major cause of the rapid rise in annual electric energy consumption in the United States.

4. The man who waters the workers' beer.
   The owner of a bar keeps his beer in a 60-gallon barrel. Each night the bartender serves the workers 6 gallons. Each morning the owner adds 6 gallons of water to the barrel.
   (a) Find a formula showing how the concentration of beer in the barrel depends on the number of days since the barrel was bought. What happens in the long run?
   (b) The workers notice when their drink is less than 60% beer. When does this happen?
   (c) Suppose that the owner starts tending bar himself. Rather than wait for the morning, he replaces each stein of beer with a stein of water immediately. There are 15 steins in a gallon. Find a formula showing how the concentration of beer in the barrel depends on the number of days since the barrel was tapped. (*Hint*: This problem resembles the periodic compounding of interest in Example 1 of Chapter 13.) When does the barrel contain less than 60% beer?
   (d) Suppose that the barrel is secretly connected to the water supply, so that as beer is drawn it is continuously replaced by water. Find a formula showing how the concentration of beer in the barrel depends on the number of days since the barrel was tapped. (*Hint*: This problem resembles the continuous compounding of interest studied in Example 2 of Chapter 13.) When does the barrel contain less than 60% beer?

5. Sour milk.
   Look again at Table 3.1 on page 43, which shows how the length of time it takes milk to sour depends on the temperature at which it is stored.
   (a) Draw a graph exhibiting the data.
   (b) Interpolate exponentially to estimate the life of the milk at 36°F and at 53°F. Extrapolate exponentially to estimate its life at 80°F, at 90°F, and at 0°F.
   (c) Compare your answers to the corresponding linear estimates. (See Chapter 3, Problem 11.) Which seem more reliable?
   (d) Play around until you find a single exponential curve of the form

$$\text{keeping time} = Ae^{-kT}$$

   that fits the data as well as possible. Draw the graph.

(e) Try to improve the fit you found in (d) by looking for two exponential curves, one to match the data for temperatures below 50°F, the other for temperatures above 50°F.

6. Postal rates.
Table 15.1 gives a history of the cost of mailing first-class a one-ounce letter in the United States over the past century.
(a) Draw a graph displaying this information.
(b) Find the annual growth factor corresponding to each increase in postal rates.
(c) Is the growth exponential? Nearly so? Predict the time of the next increase and its amount.

Table 15.1 Postal rates

Date	New rate (cents)
July 1, 1885	2
Nov. 3, 1917	3
July 1, 1919	2
July 6, 1932	3
Aug. 1, 1958	4
July 7, 1963	5
July 7, 1968	6
May 16, 1971	8
March 2, 1974	10
Dec. 31, 1975	13
May 29, 1978	15
March 22, 1981	18
Nov. 1, 1981	20

7. Recognizing qualitative behavior.
The titles have been omitted from the eight graphs in Figure 15.1. Please choose correct captions from those which follow. Name the variables. When possible, put reasonable scales on the axes. Notice that there are two more graphs than captions; make up captions to go with them.
(a) How the effective annual growth factor in continuous compounding depends on nominal annual rate.
(b) The logarithm curve.
(c) How doubling time (in exponential growth) depends on the coefficient in the exponent to which e is raised.
(d) An exponential model for depreciation.

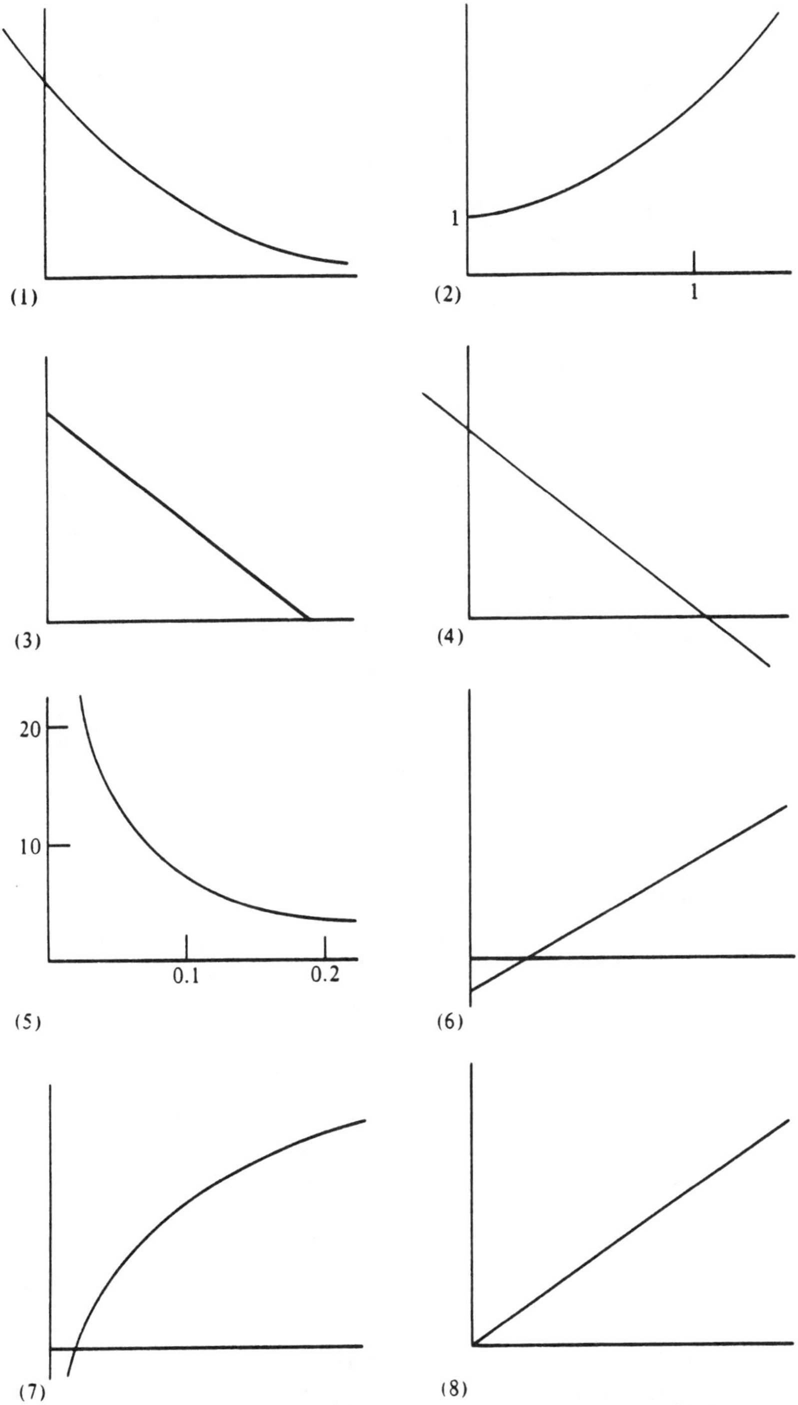

Figure 15.1 Eight unlabeled graphs

(e) A linear model for depreciation.

(f) How the logarithm of the amount of a radioactive substance depends on time.

8. An epidemic.
An epidemic is spreading. T months after its start, P people are sick. The dependence of P on T is given by

$$P = 2000\,Te^{-0.3T}.$$

(a) Draw a graph showing how P varies for the first two years. (Be sure to find $P(0)$.)

(b) When is the number of affected individuals at a maximum? What happens as the years go by?

(c) Find, by trial and error, the times at which 1000 people are ill.

9. Faster than a speeding bullet.
The time it takes light to travel short distances–a few miles–is negligible compared to longer measurable times.

(a) A physicist on a hill flashes a light at a second physicist on another hill D miles away. When the second physicist sees the first one's signal she signals in return. The time T between the first physicist's flash and his sighting of the answering light depends on the distance D, on the reaction time R of the second physicist, and on the speed of light c. Write the equation showing this dependence.

(b) The constant R in (a) is on the order of 100 milliseconds. Explain why, when physicists tried to calculate c by measuring R, T, and D, they met with no success.

(c) Sound travels at about 1000 ft/sec. If you see a flash of lightning and count seconds until you hear the thunder and divide by 5, you will know the distance in miles to the storm. Explain why.

10. Permutations.
Find as many of these as you can without your calculator. Which are (nearly) impossible even with it?

(a) $10^{\ln(e^{\log 1000})}$ (b) $10^{\log(e^{\ln 1000})}$

(c) $10^{\log[\ln(e^{1000})]}$ (d) $10^{(e^{\ln(\log 1000)})}$

(e) $10^{(e^{\log(\ln 1000)})}$ (f) $10^{\ln[\log(e^{1000})]}$

11. Some very large numbers.
    Solve for Q, R, and S.
    (a) $\log Q = 3$ (b) $\log(\log R) = 3$
    (c) $\log[\log(\log S)] = 3$

12. Solve $Z - 3 = \log Z$. There are two solutions. You may need to use trial and error to find them.

13. How many were going to St. Ives? See the rhyme that opens this chapter.

Part III

Quadratic Models

CHAPTER 16

Quadratic Variation

My purpose is to set forth a very new science dealing with a very ancient subject. There is, in nature, perhaps nothing older than motion, concerning which the books written by philosophers are neither few nor small; nevertheless I have discovered by experiment some properties of it which are worth knowing and which have hitherto not been either observed or demonstrated. Some superficial observations have been made, as, for instance, that the free motion of a heavy falling body is continuously accelerated; but to just what extent this acceleration occurs has not yet been announced; for so far as I know, no one has yet pointed out that the distances traversed, during equal intervals of time, by a body falling from rest, stand to one another in the same ratio as the odd numbers beginning with unity.

– Galileo Galilei, *Dialogues Concerning Two New Sciences*

Free fall / When one variable depends on the square of another / Kepler's and other power laws / Log-log graphs

EXAMPLE 1. FREE FALL

When a stone is dropped from the top of New York's Empire State Building the distance it falls, S, measured in feet, depends on the length of time it falls. We measure the time t in seconds. Galileo performed experiments (some say from Pisa's leaning tower) which showed that S is related to t by the formula

$$S = 16t^2. \tag{16.1}$$

What is S when $t = 2.7$ seconds? Draw the graph of Equation 16.1. Solve the equation for t in terms of S. How long does it take the stone to hit the sidewalk 1250 feet below?

Substituting $t = 2.7$ in Equation 16.1 gives

$$S = 16 * 2.7^2 = 117 \text{ ft},$$

which answers the first question. To answer the second, use Table 16.1 to draw the graph in Figure 16.1. The graph seems at first to resemble the ones we are familar with for exponential growth. It starts rising slowly, then climbs more and more steeply. But it differs in several significant ways. It begins with $S = 0$ when $t = 0$, unlike an exponential graph, which is never 0. And although it grows, it grows much more slowly than an exponential graph.

The constant 16 in Equation 16.1 is not an intercept, nor is it a slope or a growth factor. It is a proportionality constant. In this example S is proportional to t^2, not to t; and so the units of the proportionality constant 16 must be

$$\frac{\text{units of } S}{(\text{units of } t)^2}$$

or

$$\text{ft/sec}^2.$$

We will find out what these units mean in Chapter 17.

Table 16.1 Free fall

Time spent falling t (seconds)	Distance covered S (feet)
0	0
0.5	4
1	16
1.5	36
2	64
2.5	100
2.7	117
3	144
4	256
5	400
6	576
7	784
8	1024
9	1296

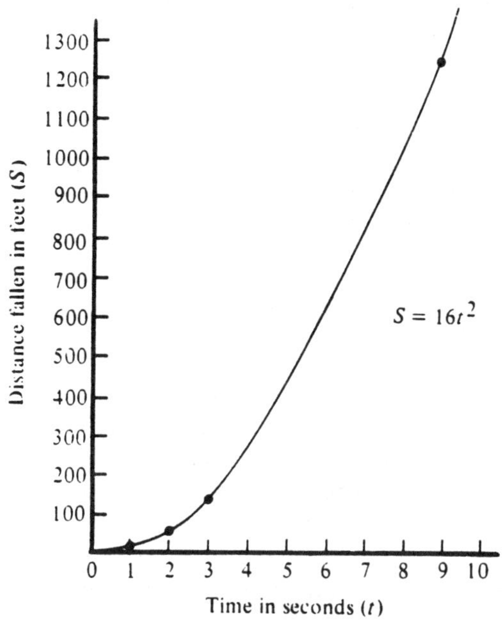

Figure 16.1 Free fall

Next we must solve Equation 16.1 for t in terms of S. If we divide each side by 16, we get

$$t^2 = \frac{S}{16} = 0.0625S.$$

Extracting the square root of each side gives:

$$\begin{aligned} t &= \sqrt{\frac{S}{16}} \quad \text{or} \quad \sqrt{0.0625S} \\ &= \frac{\sqrt{S}}{4} \quad \text{or} \quad 0.25\sqrt{S} \end{aligned} \tag{16.2}$$

Here we have used the facts that

$$\sqrt{AB} = \sqrt{A}\sqrt{B} \tag{16.3}$$

and

$$\sqrt{\frac{A}{B}} = \frac{\sqrt{A}}{\sqrt{B}}. \tag{16.4}$$

To find out how long it takes the stone to fall 1250 ft, we substitute $S = 1250$ in Equation 16.2 and conclude that

$$t = 8.84 \text{ seconds.}$$

It takes nearly 9 seconds for the stone to fall 1250 feet. This answer is consistent with Table 16.1 and with the graph in Figure 16.1.

EXAMPLE 2. JAMMING ON THE BRAKES

The distance it takes a car to stop is proportional to the square of the car's velocity when the brakes are first applied. A Chevrolet involved in an accident left skid marks 240 feet long. The police know that this car needs 170 feet to stop from a velocity of 60 miles/hour. How fast was it traveling just before the accident? If the speed of a car is doubled, what happens to the distance it takes to stop it?

The variables are D, for stopping distance, and V, for the velocity. We are told that

$$D = k * V^2 \tag{16.5}$$

for some constant k. Since D is measured in feet and V in mi/hr, the units of k are ft/(mi/hr)^2. To find the value of the proportionality constant k, we substitute $D = 170$ when $V = 60$ in Equation 16.5, giving

$$170 = k * 60^2$$

Then we solve this for k:

$$3600k = 170,$$

and so

$$k = \frac{170}{3600} = 0.047 \frac{\text{ft}}{(\text{mi/hr})^2}.$$

Equation 16.5 now becomes

$$D = 0.047 * V^2. \tag{16.6}$$

To find how fast the car was going we infer from the skid marks that

$D = 240$ ft. Substituting in Equation 16.6 gives

$$240 = 0.047V^2,$$

so

$$V = \sqrt{240/0.047} = 71 \text{ mi/hr}.$$

No wonder there was an accident!

The last question of this example asks what happens to D when the velocity doubles. If D were proportional to V, it would double too. But D is proportional to the *square* of V. For example, suppose the car is traveling 30 mi/hr. Then the stopping distance is

$$D(30) = k * 30^2 = 900k \text{ ft}. \qquad (16.7)$$

If the speed doubles to 60 mi/hr, the stopping distance is

$$D(60) = k * 60^2 = 3600k \text{ ft}. \qquad (16.8)$$

If we compare Equations 16.7 and 16.8, we see that

$$D(60) = 4 * D(30).$$

When you travel twice as fast it takes four times as far to stop.

EXAMPLE 3. RESCUE AT SEA

Lifeboats carry an orange dye, which spreads out on the surface of the sea to make a patch visible to searchers in airplanes. The radius of the circle formed by the dye is proportional to the square root of the amount of dye. Tests show that 0.5 liter of dye covers a circle of radius 1.6 m. How large a circle will 4.2 liters of dye make?

Let R be the radius of the circle covered by D liters of dye. We are told that

$$R = k\sqrt{D}. \qquad (16.9)$$

We know from the data given that

$$1.6 = k\sqrt{0.5},$$

so

$$k = \frac{1.6}{\sqrt{0.5}} = 2.3 \text{ m}/\sqrt{\text{L}}.$$

Therefore, when $D = 4.2$,

$$R = 2.3\sqrt{4.2} = 4.6 \text{ m}.$$

The radius R is proportional to the square root of D because the dye spreads on the surface in a layer of uniform thickness. Thus the area covered is proportional to the volume of dye. But the area is proportional to the square of the radius. (The proportionality constant is π.) Thus R^2 is proportional to D, or:

$$D = KR^2 \tag{16.10}$$

for some constant K. When we take the square root of each side of Equation 16.10 and set

$$k = \sqrt{K},$$

we get Equation 16.9. Thus

$$K = k^2 = 5.1 \text{ m}^2/\text{L}.$$

EXAMPLE 4. INVERSE SQUARE LAWS

One of Newton's most important discoveries was the law that describes how the gravitational force between two objects depends on the distance that separates their centers. Gravitational force is *inversely proportional* to the *square* of the distance. If an astronaut weighs 135 pounds at the surface of the earth, what will she weigh at an altitude of 18 000 miles? At the top of the Empire State Building?

The two variables in this problem are the astronaut's weight, W, measured in pounds, and the distance from the center of the earth to the (center of the) astronaut. We measure this distance r in miles. Newton's law is expressed by the equation

$$W = \frac{K}{r^2}. \tag{16.11}$$

At the surface of the earth, $r = 4000$ miles and $W = 135$ pounds. We use these figures to find K.

$$135 = \frac{K}{4000^2},$$

so

$$K = 135 * 4000^2 = 2.16 * 10^9.$$

The units of K are lb * mi². Thus,

$$W = \frac{2.16 * 10^9}{r^2} \text{ lb.} \qquad (16.12)$$

At an altitude of 18 000 miles the distance to the center of the earth is $r = 22\,000$ miles, and

$$W = \frac{2.16 * 10^9}{22\,000^2} = 4.5 \text{ pounds.}$$

To find W when the astronaut stands at the top of the Empire State Building we need to know the value of r there. It is the distance to the center of the earth, or

$$4000 \text{ miles} + 1250 \text{ feet} \sim 4000.2 \text{ miles.}$$

Then

$$W = \frac{2.16 * 10^9}{4000.2^2} = 134.987 \text{ pounds.}$$

This differs negligibly from the 135 pounds the astronaut weighs at the foot of the building. Newton's law of gravitation predicts major changes in weight only when there are appreciable changes in the distance r. Though 1250 feet may seem like a sizable distance, it is small compared to the 4000-mile value of r at the earth's surface. Over such small distances the force of gravity is virtually constant. This is why Galileo's experiments discussed in Example 1 gave the results they did.

We have just found that at a height of 18 000 miles the astronaut weighs 4.5 pounds. This seems to contradict what we read about the weightlessness of astronauts in space. It doesn't. Our computation of the force of gravity is accurate. It is the speed of a satellite in its orbit that causes the phenomenon of weightlessness. Were it not for gravity, that speed would carry the satellite away from the earth in a straight line. Gravity bends the path. The satellite and the astronauts within fall freely toward the earth at a rate exactly matching the rate at which their orbital speed would carry them away.

One further point: This problem would look different in the metric system. At the surface of the earth the astronaut's 135 pounds converts to

$$135 \text{ pounds} * 0.455 \frac{\text{kg}}{\text{pound}} = 61.4 \text{ kg}.$$

At an altitude of 18 000 miles she weighs only 4.5 pounds, but there is *still* 61 kg of astronaut. This paradox is explained by a subtle distinction between *mass* and *weight*. *Mass* is how much of something there is; *weight* is how hard gravity pulls on it. The pound is really a unit of weight; the kilogram is a unit of mass. An astronaut in orbit has the *same mass* as on the earth's surface, but weighs a lot less.

EXAMPLE 5. POWER LAWS

Kepler's third law of planetary motion says that the square of the period T of a satellite or planet is proportional to the cube of the radius r of its orbit:

$$T^2 = kr^3, \tag{16.13}$$

where k is some constant. The moon is 240 000 miles from the earth and has a period of 28.3 days. Find the period of a satellite whose altitude above the earth is 10 000 miles. At what altitude should a satellite orbit so that it will hover constantly over a fixed point on the earth? Graph the dependence of T on r and of $\log(T)$ on $\log(r)$.

As usual, we begin by finding the proportionality constant. Substituting the data into Equation 16.13 gives

$$28.3^2 = k * 240\,000^3,$$

so

$$k = \frac{28.3^2}{240\,000^3} = 5.8 * 10^{-14} \frac{\text{day}^2}{\text{mile}^3}.$$

A satellite orbiting at a height of 10 000 miles has an orbital radius of 14 000 miles. Its period T is given by

$$T = \sqrt{5.8 * 10^{-14} * 14\,000^3} = 0.40 \text{ days},$$

or about 9.6 hours.

A geosynchronous satellite—one that hovers over a single spot (and is therefore useful for communication)—has a period T exactly equal to one day. For such a satellite

$$1 = T^2 = 5.8 * 10^{-14}\, r^3,$$

so

$$r = \left(\frac{1}{5.8 * 10^{-14}}\right)^{1/3} = 26\,000 \text{ miles}.$$

(Remember that raising a number to the ⅓ power is the same as finding its cube root.) Then 26 000 miles from the earth's center corresponds to an altitude of 22 000 miles.

Table 16.2 shows the data we have calculated so far, together with their logarithms, plus a fourth row corresponding to T = 96 minutes. This is the period of the first man-made satellite launched in the late fifties, which orbited at an average altitude of only 250 miles. These data are used to draw the graphs in Figures 16.2 and 16.3.

That the graph in Figure 16.3 is a straight line may surprise you. To see why it is a line, take the logarithm of each side of Equation 16.13:

$$2 \log(T) = \log(k) + 3 \log(r).$$

Therefore,

$$\log(T) = \tfrac{1}{2} \log(k) + \tfrac{3}{2} \log(r).$$

Table 16.2 Orbital periods, radii, and their logarithms

T (days)	r (miles)	$\log T$	$\log r$
28.3	240 000	1.45	5.30
1	26 000	0	4.41
0.40	14 000	−0.40	4.15
0.067	4 250	−1.18	3.63

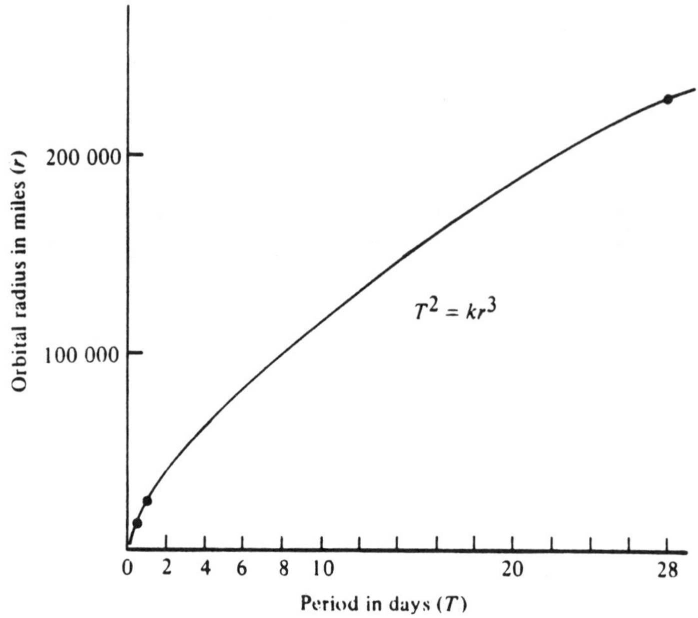

Figure 16.2 Satellite periods

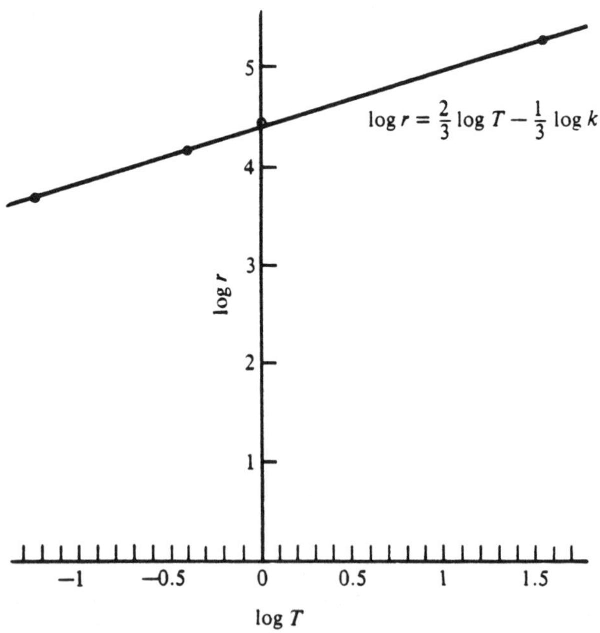

Figure 16.3 A linear log-log graph

This tells us that log(t) depends *linearly* on log(r). The slope of the line is 3/2 and the intercept is log(k)/2.

This kind of linear dependence between logarithms always happens when two variables P and Q are related by a *power law*—a proportionality between some power of Q and some power of P. If

$$Q^a = kP^b,$$

then

$$a \log(Q) = \log(k) + b \log(P)$$

so

$$\log(Q) = \frac{\log(k)}{a} + \left(\frac{b}{a}\right) \log(P).$$

PROBLEMS

1. Verify Equations 16.3 and 16.4 first numerically, by substituting some values for A and B and using your calculator, and then algebraically, by squaring each side.
2. Does $\sqrt{A+B} = \sqrt{A} + \sqrt{B}$?
   Substitute numerical values for A and B and check.

*3. Convert the constant 32 ft/sec^2 to cm/s^2 and to m/h^2.

4. Galileo's observations.
   (a) How far does a freely falling object fall from rest in 6 seconds? How far does it fall in the sixth second—between $t = 5$ and $t = 6$ seconds?
   (b) Answer the same question for the first, second, third, fourth, and fifth seconds. Prepare a table and draw a graph showing how the distance fallen in one second depends on which second is being considered.
   (c) Discuss the claim made by Galileo in the quotation that opens chapter.
5. (a) Draw graphs of the following equations:

$$Q = t^2 \quad \text{and} \quad Q = 2^t,$$

using the same set of coordinate axes. Which is larger when t is small? Where do they cross? Which grows faster when t is large? What happens to each when t is negative? Be sure your graphs show this information too.

(b) Draw graphs showing how $\log Q$ depends on t in each case.

*6. It takes a Ford Fairmont 175 feet to stop when it is traveling 60 miles per hour. Find the stopping distance when it is traveling 30, 50, and 90 miles per hour, by finding a formula that expresses stopping distance in terms of velocity. Draw a graph illustrating this dependence. Indicate the units of all variables and constants you use.

7. Seven pounds of grass seed will cover a circular lawn 25 feet in diameter. How much grass seed is needed for a lawn 50 feet in diameter? For a lawn 55 feet in diameter? For a rectangular lawn 30 by 45 feet?

8. (a) Explain why the radius of a soap bubble is proportional to the square root of the volume of soap solution used to blow the bubble. (*Hint*: Look up the formula for the surface area of a sphere.)

(b) If 1.3 mL of solution forms a bubble 5 cm in diameter, how much will be required for a bubble 10 cm in diameter?

9. (a) Draw a graph illustrating how the weight of the 135-pound astronaut (in Example 4) depends on her distance above the surface of the earth.

(b) At what altitude will she weigh 100 pounds?

(c) At what altitude will she weigh half what she weighs at the earth's surface? At what altitude is her weight 1/4 of what it was?

(d) How do your answers in (b) and (c) resemble those you would get from a model of exponential decay? How do they differ?

*10. (a) Solve Equation 16.13 for T in terms of r to get a formula of the form:

$$T = Cr^A.$$

What is the numerical value of the constant C? What are its units?

(b) Solve Equation 16.13 for r in terms of T to get a formula of the form:

$$r = cT^a.$$

What is the numerical value of the constant c? What are its units?

11. Table 16.3 shows the length of the year in earth days and the distance from the sun in kilometers for each of the planets. Draw a graph showing how the logarithms of these numbers are related. Does your result confirm Kepler's law? (He might have guessed the law from this kind of graph. He knew about the planets as far out as Saturn.) What is the value of the proportionality constant?

Table 16.3 Data about the solar system

Planet	Length of year (days)	Radius of orbit (kilometers)
Mercury	88	5.8 * 10^7
Venus	225	1.08 * 10^8
Earth	365	1.49 * 10^8
Mars	687	2.28 * 10^8
Jupiter	4 333	7.78 * 10^8
Saturn	10 759	1.43 * 10^9
Uranus	30 686	2.87 * 10^9
Neptune	60 188	4.50 * 10^9
Pluto	90 885	5.90 * 10^9

*12. If Q is proportional to R^p, what is the effect on Q of doubling R?

*13. Suppose that $V = kP^c$ where k and c are constants.
If $V = 16.33$ when $P = 3.21$, and $V = 102.7$ when $P = 11.4$, find k and c. (*Hint*: Take logarithms, then solve two linear equations simultaneously.)

14. Suppose a cubical box has a side that is L cm long.
   (a) What is the volume of the box in cm^3?
   (b) What is the surface area of the box in cm^2?
   (c) Show how the ratio of the area of the box to its volume decreases as L increases.

The fact that the surface-to-volume ratio decreases as the length increases has important consequences in biology. For example, the rate at which an organism can take up nutrients depends on the area of its digestive tract, but the amount of nourishment it needs depends on its volume. Thus if an animal were suddenly to double in size, its surface-to-volume ratio would decrease (to half its former value) and it would no longer be able to absorb nutrients fast enough. That is why large animals have highly twisted digestive tracts, with large areas in proportion to their size.

15. Stefan's law states that the rate at which a hot body radiates energy is directly proportional to the fourth power of its absolute (or Kelvin) temperature. The absolute temperature scale has its zero point at $-273°$ Celsius; and the size of the degrees is the same on the two scales. On the absolute scale the units are called *kelvins* (K). Thus,

$$0°\text{C} = 273\text{ K} \quad \text{and} \quad 20°\text{C} = 293\text{ K}.$$

When the temperature is measured in kelvins, energy in Btu's (British thermal units), area in square feet, and time in hours, the constant in Stefan's law has the value

$$4.2 * 10^{-9} \frac{\text{Btu}}{\text{ft}^2 * \text{hr} * \text{K}^4}.$$

How much heat is given off by a radiator whose area is 36 square feet when its temperature is 100°F? (First convert Fahrenheit to Celsius degrees, then to kelvins.)
(*Note*: Real room-heating radiators do not actually obey this simplified version of Stefan's law.)

16. Calculate the value of each of the following and express your answers in scientific notation:

(a) $8.7 * 10^{12}$	(b) 1.6	(c) $9.99 * 10^{44}$
(d) $1.32 * 10^{-18}$	(e) $7.35 * 10^{112}$	(f) $1 * 10^{56}$
(g) $9 * 10^{2N}$	(h) $4 * 10^{-256}$	(i) $(3.1 * 10^5)^2$
(j) $(3.6 * 10^5)^2$	(k) $(3.1 * 10^N)^2$	(l) $(3.6 * 10^N)^2$

(m) Use the results from (a)–(l) to make up a rule for finding the square root of a number that, when written in scientific notation, shows 10 raised to an even power. How large can the coefficient be in the answer?

CHAPTER 17

Quadratic Models

How doth the falling article
Describe its curving trail,
In figure parabolical–
Ly perfect, without fail?

Through forces gravitational
It always finds the ground;
I think these routes equational
Must be the long way 'round.

– Jessica Bolker

Velocity and acceleration / High and low points and roots in quadratic models / The quadratic formula

EXAMPLE 1. FREE FALL REVISITED–CONSTANT ACCELERATION

Suppose we drop another stone from the Empire State Building. In t seconds it falls

$$S = 16t^2 \text{ feet.} \tag{17.1}$$

How fast is it falling after 3 seconds? After 3.7 seconds? Explain how its speed changes with time.

What makes this question both interesting and difficult is that the speed V, measured in ft/sec, is changing. The stone starts out falling slowly and gains speed as time goes by. If the distance S depended linearly on the time t, the speed V would be the constant slope of the line, and we could find it by calculating the ratio

$$\frac{\text{change in } S}{\text{change in } t}. \tag{17.2}$$

But since the dependence is not linear, this ratio is not constant. For example, between times $t = 3$ sec and $t = 3.5$ sec we have

$$\begin{aligned} \text{average speed} &= \frac{\text{change in } S}{\text{change in } t} \\ &= \frac{S(3.5) - S(3)}{3.5 - 3} \\ &= \frac{16(3.5^2) - 16(3^2)}{3.5 - 3} \\ &= 104 \text{ ft/sec,} \end{aligned} \tag{17.3}$$

whereas between times $t = 3$ sec and $t = 3.1$ sec,

$$\begin{aligned} \text{average speed} &= \frac{\text{change in } S}{\text{change in } t} \\ &= \frac{S(3.1) - S(3)}{3.1 - 3} \\ &= 97.6 \text{ ft/sec.} \end{aligned} \tag{17.4}$$

Of these two calculations the second is likely to be closer to the actual speed when $t = 3$ sec, because the time interval over which the average speed was calculated is shorter. To find an even better approximation to the actual speed when $t = 3$ sec, find the average speed between $t = 3$ sec and $t = 3.001$ sec:

$$\begin{aligned} \text{average speed} &= \frac{\text{change in } S}{\text{change in } t} \\ &= \frac{S(3.001) - S(3)}{3.001 - 3} \\ &= 96.016 \text{ ft/sec.} \end{aligned} \tag{17.5}$$

It looks very much as if

$$V(\text{at 3 sec}) = 96 \text{ ft/sec.} \tag{17.6}$$

To verify this, and to find the speed at any time (not just when $t = 3$), we shall carry out the kind of calculations we made in Equations 17.3, 17.4, and 17.5 algebraically rather than numerically.

In order to minimize the effect of the curvature when calculating the slope,

$$\frac{\text{change in } S}{\text{change in } t},$$

we use two times, p and q, which are very close together. Then,

$$\text{change in } t = (q - p) \text{ sec},$$

while

$$\text{change in } S = 16\,q^2 - 16\,p^2 \text{ ft}.$$

Thus between times p and q the average speed is given by the slope

$$\begin{aligned} \frac{\text{change in } S}{\text{change in } t} &= \frac{16\,q^2 - 16\,p^2}{q-p} \\ &= \frac{16(q+p)\cancel{(q-p)}}{\cancel{q-p}} \qquad (17.7) \\ &= 16(q+p). \end{aligned}$$

(Equation 17.7 follows from the factoring identity

$$A^2 - B^2 = (A+B)(A-B).$$

that you learned in and may remember from an earlier algebra course.)

You need not master the steps leading to Equation 17.7, but should think about their overall meaning: We now know how to compute the average speed during *any* time interval. To check the result, we can use it to verify Equations 17.3, 17.4, and 17.5.

Now suppose we wish to compute the speed V when $t = 3.7$ seconds. We use Equation 17.7 for two values of p and q, each very near 3.7. Then $p + q$ will be near $2 * 3.7$, and

$$V(3.7) = 16 * 2 * 3.7 = 32 * 3.7 = 118.4 \text{ ft/sec}.$$

In general, if p and q are both very near t, then the average speed

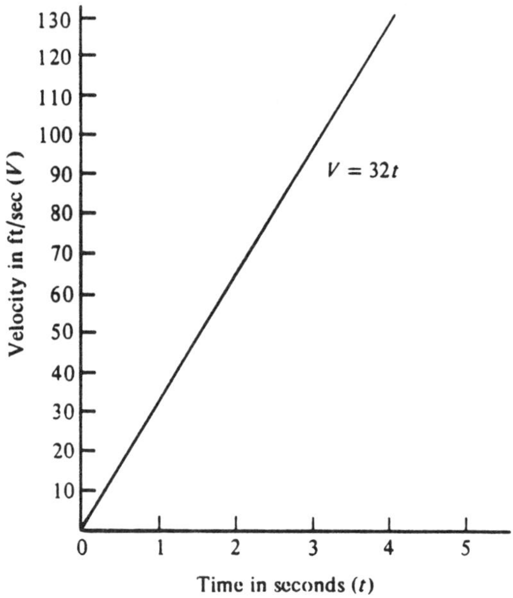

Figure 17.1 Velocity in free fall

between these times will be very near the actual speed V at time t. Moreover, $p + q$ will be near $2t$, and so Equation 17.7 tells us that

$$V(t) = 32t. \tag{17.8}$$

We knew all along that V increased as t did, since the stone fell faster and faster. Now we see exactly how rapidly V grows. Equation 17.8 describes a direct proportion. It says that V increases at a constant rate as t increases. Figure 17.1 is a graph of the dependence of V on t.

The speed V starts at 0 ft/sec and increases by 32 ft/sec each second. The rate at which the velocity changes is called the *acceleration*. The units of acceleration are:

$$\frac{\text{units of } V}{\text{units of } t} = \frac{\frac{\text{ft}}{\text{sec}}}{\text{sec}}$$

This complex fraction expressing the units of the acceleration is commonly written as

$$\frac{\text{ft/sec}}{\text{sec}} = \frac{\text{ft}}{\text{sec}} * \frac{1}{\text{sec}} = \frac{\text{ft}}{\text{sec}^2}.$$

We encountered this form in Equation 16.1.

What Galileo discovered about freely falling bodies is that near the earth's surface their acceleration—the rate at which their speed changes—is a constant 32 (ft/sec)/sec. If the effect of air resistance is ignored this constant acceleration does not change as the body falls, and is independent of the body. Big stones and small stones fall with the same constant acceleration.

Constant acceleration is the hallmark of quadratic variation. Whenever

$$Q = k * t^2,$$

the rate at which Q changes, which we can call V, is itself changing at a constant rate. The argument in Equation 17.7 can be repeated to show that

$$V = 2kt.$$

The units of the rate V are

$$\frac{\text{units of } Q}{\text{units of } t},$$

and so the units of constant acceleration $2k$ are

$$\frac{\text{units of } V}{\text{units of } t} = \frac{\dfrac{\text{units of } Q}{\text{units of } t}}{\text{units of } t} = \frac{\text{units of } Q}{(\text{units of } t)^2}. \tag{17.9}$$

The acceleration due to gravity, which is what makes objects fall, is

$$\frac{32 \text{ ft/sec}}{\text{sec}} \quad \text{or} \quad 32 \frac{\text{ft}}{\text{sec}^2}.$$

EXAMPLE 2. MEASURING THE DISTANCE FROM THE GROUND

What formula tells us how far a falling stone is from the ground at any time? When does it strike the ground? What is its speed at any time during its fall?

We know that t seconds after being released the stone has fallen

$$S = 16\, t^2 \text{ ft}.$$

Thus when it falls from the top of the Empire State Building its height H above the ground is

$$H = 1250 - 16t^2 \text{ ft.} \tag{17.10}$$

Figure 17.2, the graph of this equation, is easily constructed from Table 16.1. It is the graph of Figure 16.2, turned upside down and raised 1250 feet. This is because, when $t = 0$ seconds, $H = 1250$ feet. Thus 1250 feet is the intercept.

We know how to find what value of t gives $H = 0$; it is the point where the graph in Figure 17.2 crosses the horizontal axis. If we substitute $H = 0$ in Equation 17.10 and solve for t, the result is

$$t = \frac{1250}{16} = 8.84 \text{ seconds.} \tag{17.11}$$

The velocity V at time t is given by

$$V = -32t. \tag{17.12}$$

The negative sign deserves a comment. In Example 1 we measured

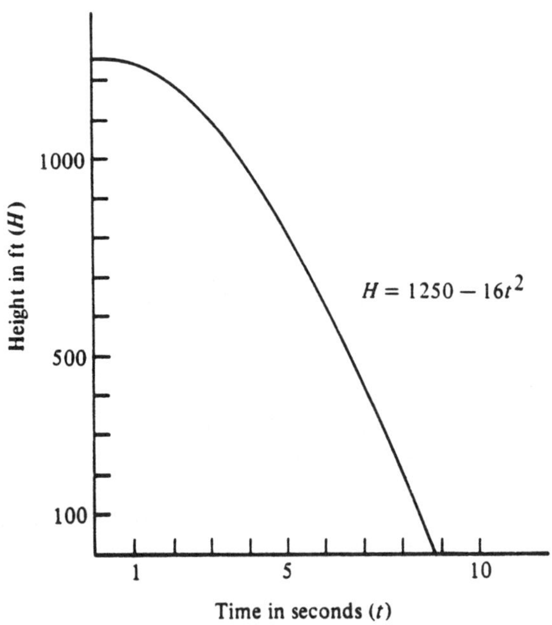

Figure 17.2 Path of a falling stone

the distance the stone had fallen. This distance increased faster and faster as time increased, so that the velocity, which measured the rate at which the distance increased, grew from 0 to larger and larger positive numbers. This follows from Equation 17.8. In this example, however, H starts out at 1250 ft when $t = 0$ and becomes smaller as time goes on. Thus V, which measures the rate at which H changes, must be negative. Moreover, H shrinks at a faster and faster rate: the curve in Figure 17.2 becomes steeper as you move to the right. Thus the magnitude of the negative number must increase with time. All this is conveyed by Equation 17.12.

In general, if a quantity Q depends on t according to the formula

$$Q = c + at^2, \tag{17.13}$$

where c and a are constants, then the graph showing the dependence of Q on t resembles Figure 17.2. How close the resemblance is depends on the signs of c and a. There are four possibilities: c and a are both positive, c and a are both negative, c is positive and a negative, c is negative and a positive. These are illustrated in Figure 17.3.

In the example we have just considered, $c = 1250 > 0$, and $a = -16 < 0$; Figure 17.3(c) most nearly resembles Figure 17.2. In each case c is the Q-intercept; that is, the value of Q when $t = 0$. Notice further that Equation 17.13 makes sense even when t is negative. These parts of the curves in Figure 17.3 are drawn with dashed lines. The dashed parts are mirror images of the solid parts. The complete curve is called a *parabola*.

Next we look for the t-intercepts–the values of t that make $Q = 0$. In Figures 17.3 (a) and (b) there are none, in each of (c) and (d) there are two. They can be found by setting $Q = 0$ in Equation 17.13 and solving for t:

$$0 = c + at^2,$$

so

$$at^2 = -c,$$

$$t^2 = \frac{-c}{a},$$

and

$$t = \sqrt{\frac{-c}{a}}. \tag{17.14}$$

Equation 17.11 is just a special case of Equation 17.14.

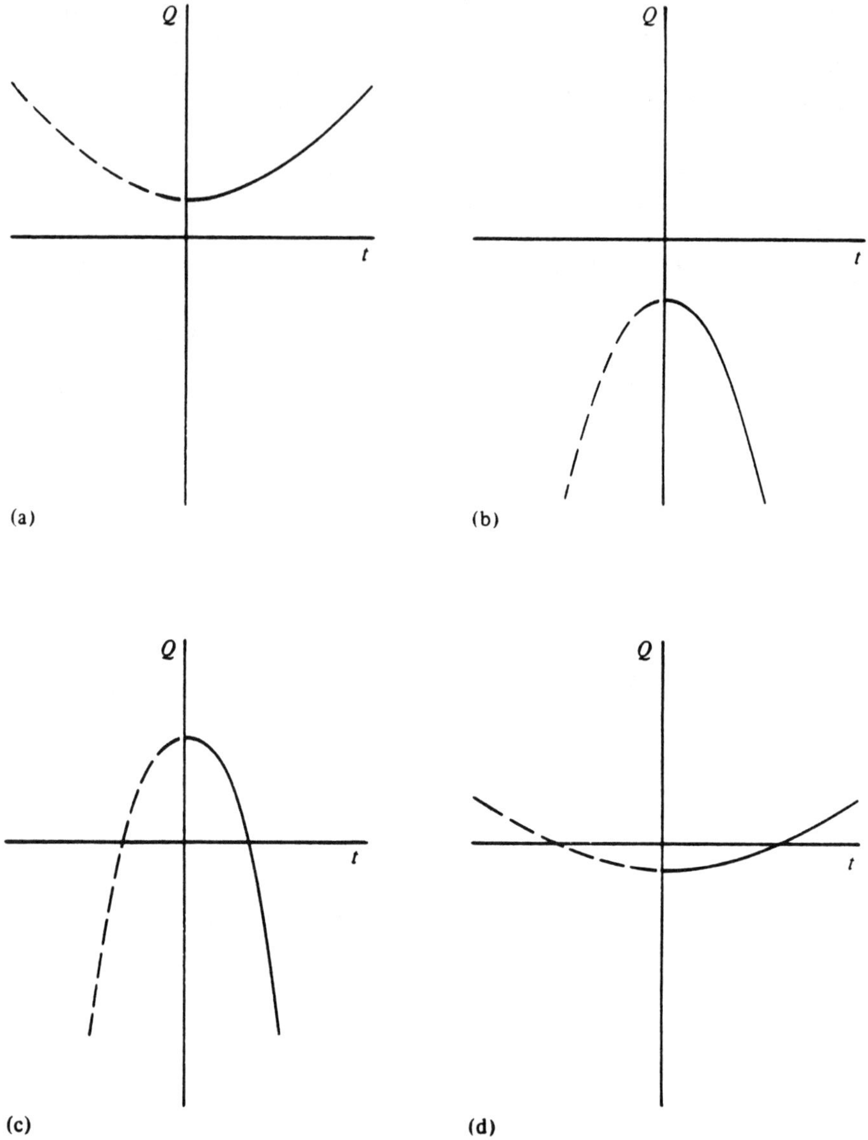

Figure 17.3 Four parabolas

When c and a have the same sign, Equation 17.14 requires you to find the square root of a negative number. But this is impossible. We have here the algebraic counterpart of the geometric fact that the parabolas in Figure 17.3(a) and (b) do not cross the t-axis. When c and a have opposite signs, $-c/a$ is positive and has two square roots.

They correspond to the two places where the parabolas ir (c) and (d) meet the t-axis.

The rate V at which Q changes is given by

$$V = 2at. \tag{17.15}$$

EXAMPLE 3. A THROWN STONE

Suppose we are standing on a bridge 100 feet high and throw a stone upward at a speed of 30 ft/sec. How high does it rise before it falls back? When does it hit the water below?

The three variables in this problem are t, the time in seconds, S, the stone's distance in feet from the water, and V, the stone's velocity in ft/sec. We know that

$$S(0) = 100 \text{ ft}$$

(since the stone starts at the bridge), and

$$V(0) = 30 \text{ ft/sec}$$

(the speed with which it was thrown). If there were no gravity, V would remain constant. But gravity is at work, applying its constant acceleration to decrease the velocity:

$$V = 30 - 32t. \tag{17.16}$$

Equation 17.16 is like linear depreciation. The velocity starts at 30 ft/sec and decreases at a steady rate of 32 (ft/sec)/sec. Figure 17.4 shows graphically how V depends on t. If there were no gravity, the stone would rise from the bridge at a steady rate given by the initial velocity of 30 ft/sec, so that

$$S \text{ would} = 100 + 30t. \tag{17.17}$$

We get the correct dependence when we subtract the effect of gravity:

$$S = 100 + 30t - 16t^2. \tag{17.18}$$

Figure 17.5 graphs the dependence of S on t.

Our job now is to find from Equations 17.16 and 17.18, and from Figures 17.4 and 17.5, the answers to the two questions asked. We start

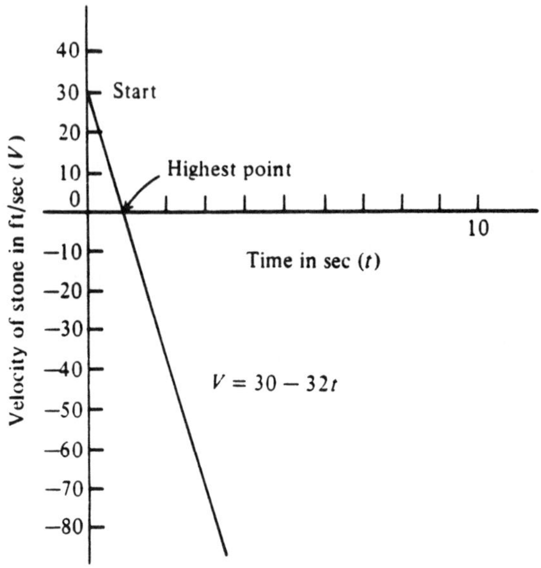

Figure 17.4 Velocity of a thrown stone

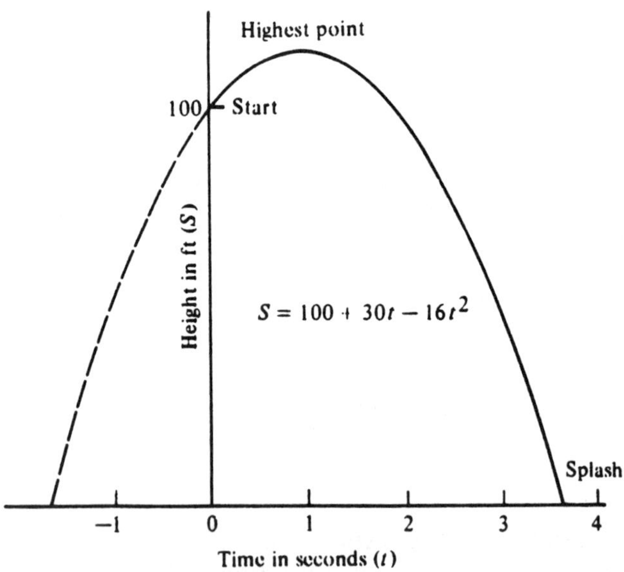

Figure 17.5 Height of a thrown stone

with the highest point. The stone rises for a while and then falls back. While it is rising, V is positive. While it is falling, V is negative. *At* the highest point, $V = 0$. Therefore we can find the *time* to the highest point by setting $V = 0$ in Equation 17.16 and solving for t:

$$0 = 30 - 32t, \tag{17.19}$$

so it takes

$$t = \frac{30}{32} = 0.94 \text{ sec.} \tag{17.20}$$

for the stone to reach its highest point.

We see this in Figure 17.4 as the point where the velocity V is 0. Until then, V is positive. Thereafter, it is negative. Now that we know when the highest point is reached, we can find how high it is by substituting in Equation 17.18. When $t = 0.94$ sec,

$$S = 100 + 30(0.94) - 16 * 0.94^2 = 114 \text{ ft.} \tag{17.21}$$

This high point shows clearly in Figure 17.5.

Finally, to discover when the stone splashes into the water, we make one basic observation: The stone falls *freely* from the high point. Therefore:

$$\text{splash time} = \text{time to high point} + \text{time to fall to the water.} \tag{17.22}$$

The time to the high point is given by Equation 17.20. The time it takes the stone to fall freely 114 ft from the high point is given by the analogue of Equation 17.11:

$$t = \sqrt{\frac{114}{16}} = 2.67 \text{ sec.} \tag{17.23}$$

Thus

$$\text{splash time} = 0.94 + 2.67 = 3.6 \text{ sec.} \tag{17.24}$$

We can see this in Figure 17.5 as the point where S is 0. If we wish to know how fast the stone is traveling when it hits the water, we need only substitute the time in Equation 17.16:

$$V \text{ (at splash)} = 30 - 32 * 3.6 = -85 \text{ ft/sec.} \tag{17.25}$$

We conclude this discussion by working the same example once again in a general situation. Suppose we stand on a bridge c feet high and throw a stone upward at a speed of b ft/sec. Suppose too that this all happens on a planet where the acceleration due to gravity is $-2a$ (ft/sec)/sec. (In the example we have just done, $a = -16$, $b = 30$, and $c = 100$.) In general, the velocity V is given by

$$V = b + 2at. \tag{17.26}$$

(Compare this with Equation 17.16.) The distance S from the water is given by

$$S = c + bt + at^2. \tag{17.27}$$

(Compare this with Equation 17.18.) The high point occurs when $V = 0$. Then

$$0 = b + 2at,$$

so that

$$\text{time to high point} = \frac{-b}{2a}. \tag{17.28}$$

(Compare this with Equation 17.20.) We find the height of the high point by substituting this time in Equation 17.27:

$$\begin{aligned} \text{Height of high point} &= c + b\left(\frac{-b}{2a}\right) + a\left(\frac{-b}{2a}\right)^2 \\ &= c - \frac{b^2}{2a} + a\left(\frac{b^2}{4a^2}\right) \\ &= c - \frac{b^2}{2a} + \frac{b^2}{4a} \\ &= c - \frac{b^2}{4a}. \end{aligned} \tag{17.29}$$

(Compare this with Equation 17.21.)

To find the splash time use Equation 17.22 as before. The time to the high point is given by Equation 17.28. The time it takes the stone to fall freely the $c - b^2/4a$ ft from the high point is given by the analogue of Equation 17.11:

$$
\begin{aligned}
t &= \sqrt{\frac{c - b^2/4a}{-a}}, \\
&= \sqrt{\frac{b^2 - 4ac}{4a^2}}, \left(\text{we multiplied by } \frac{-4a}{-4a}\right) \\
&= \frac{\sqrt{b^2 - 4ac}}{2a}.
\end{aligned}
\tag{17.30}
$$

(Compare this with Equation 17.23.) Therefore,

$$
\begin{aligned}
\text{time to splash} &= \frac{-b}{2a} + \frac{\sqrt{b^2 - 4ac}}{2a} \\
&= \frac{-b + \sqrt{b^2 - 4ac}}{2a}
\end{aligned}
\tag{17.31}
$$

(Compare this with Equation 17.24.)

Equation 17.31 is famous; it may be known to you as the *quadratic formula*. It solves the general quadratic equation:

$$at^2 + bt + c = 0 \tag{17.32}$$

for t in terms of a, b, and c. We have just derived it by studying how bodies fall. It is useful in a great many other contexts too.

Sometimes Equation 17.32 has no solution. This happens when the number $b^2 - 4ac$, whose square root we must find, is negative. The equations for graphs (a) and (b) in Figure 17.3, which never cross the t-axis, illustrate this situation.

Whenever Equation 17.32 has one solution it usually has two, because when $b^2 - 4ac$ is positive it will have two square roots. In the example of the thrown stone, the second solution is

$$t = 0.94 - 2.67 = -1.7 \text{ sec.} \tag{17.33}$$

You can see this in Figure 17.5 as the point where the dashed part of the curve meets the t-axis. We can infer from Equations 17.25 and 17.33 that if someone were to throw a stone upward from the river 1.7 seconds before we threw ours, and with a velocity of 85 ft/sec, his stone and ours would follow the same paths.

EXAMPLE 4. OPTIMIZATION

The firm of White & Daughters manufactures widgets. They find that their monthly profit P (measured in thousands of dollars) depends on the widget selling price s (in dollars) according to the quadratic rule:

$$P = -3.15s^2 + 16.7s - 15.2. \tag{17.34}$$

What selling price maximizes profits? What selling prices correspond to breaking even? Draw the graph showing how P depends on s.

We shall see in Problem 4 why a quadratic model provides a reasonable description of the dependence of profit on selling price. For now, let us study the consequences of Equation 17.34. We start with Table 17.1, which shows the monthly profit at various widget selling prices. We can see from the graph in Figure 17.6 that there are two selling prices that lead to breakeven–one between \$1.00 and \$2.00 (where the selling price barely covers costs), and one between \$4.00 and \$5.00. (There the selling price is reaching the point where falling sales decrease profits to 0.) To find the breakeven selling prices exactly, we can use the quadratic formula. It will tell us the roots of Equation 17.34, which are the values of s that make $P = 0$. Here $a = -3.15$, $b = 16.7$, and $c = -15.2$, so that Equation 17.34 becomes:

$$s = \frac{-16.7 \pm \sqrt{16.7^2 - 4(-3.15)(-15.2)}}{2(-3.15)} = \$4.14 \text{ and } \$1.17.$$

These are the breakeven selling prices. The selling price that maximizes the profit corresponds in this example to the time the stone reached its maximum height in the last example. Thus, the price is given by

$$s = \frac{-b}{2a} = \frac{-16.7}{2(-3.15)} = \$2.65.$$

Table 17.1 White and Daughters profit on widget sales

Selling price (\$)	Monthly profit (thousands of \$)
1	−1.65
2	5.60
3	6.55
4	1.20
5	−10.45

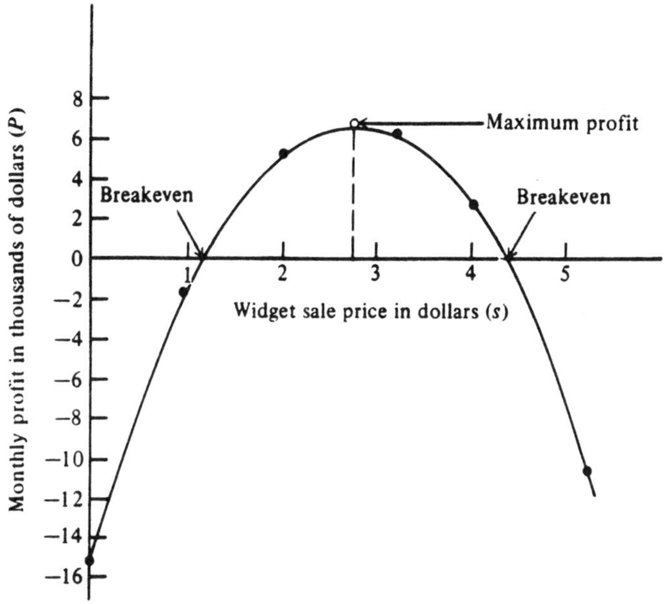

Figure 17.6 White and Daughters monthly profit

At this selling price, Equation 17.34 tells us that the monthly profits are $6930.

The data from Table 17.1, the two breakeven selling prices, and the optimal selling price can all be seen in Figure 17.6.

PROBLEMS

*1. A stone is thrown upward from a 120-foot bridge with an initial velocity of 24 ft/sec.

   (a) After 2 seconds how high above the water is the stone, and how fast is it traveling?

   (b) After how many seconds does it reach its highest point, and how high is it then?

   (c) How long is it before it hits the water, and how fast is it moving then?

2. Schoolchildren often think that $4 * 6$ should be the same as $5 * 5$, and that, in general, $(A - 1)(A + 1)$ should equal A^2.

   (a) Draw pictures of 4×6 and 5×5 rectangles and explain how you could use them to show schoolchildren why 24 doesn't equal 25.

(b) If you were good at squares, how could you compute the value of $(A - 1)(A + 1)$ rapidly?

(c) Find 96 * 104 using neither your calculator nor pencil and paper by using the formula:

$$(A - B)(A + B) = A^2 - B^2.$$

(d) Show algebraically why this formula is true.

3. A lawnmower manufacturer estimates that his daily dollar profit P depends on the number L of lawnmowers produced according to the formula:

$$P(L) = -1200 + 55L - 0.1L^2.$$

(a) Draw a graph showing how P depends on L.

(b) What is the daily overhead?

(c) At what production levels does the company break even?

(d) What production level maximizes profits?

*4. Mr. Woolf, the baker, finds that when he charges \$0.93 per loaf he sells 400 loaves of bread per week. If he raises the price to \$1.09 per loaf he sells only 250 loaves per week.

(a) Suppose that the number of loaves he can sell depends linearly on the price he charges. Write an equation showing this dependence. Identify the slope and intercept and their units.

(b) How many loaves could he sell if he charged \$0.97 per loaf?

(c) How should he set the price to sell 430 loaves?

(d) Suppose it costs him \$0.72 to make each loaf. Calculate his *profit* in weeks when he charges \$0.93, \$0.97, and \$1.09 per loaf.

(e) Find a formula showing how his weekly profit depends on the price he charges, and find out what price maximizes his profit.

(f) At what prices does he make no profit at all?

5. The Lord told Gideon his army of 32 000 was too large. Twice he dismissed a fraction of his troops. Then he was left with the 300 he was intended to have. Suppose that he sent home the same fraction f each time. Find f. (*Hint*: Explain why f satisfies the equation:

$$32\,000(1 - f - f^2) = 300.$$

Then solve the equation. As a matter of fact, Gideon's army was first reduced to 10 000 and then to 300, so that the dismissed fraction was not the same each time.)

*6. A major league pitcher's fast ball travels the 64 ft from the mound to home plate at about 90 mi/hr.

(a) How long does it take?

(b) How much will gravity make the ball drop during this time?

7. At a distance of D km from the earth the gravitational force pulling a spaceship toward the earth is

$$\frac{5.98 * 10^{27}\,k}{D^2},$$

while the force pulling the ship toward the moon is

$$\frac{7.34 * 10^{25}\,k}{(385\,000 - D)^2}.$$

At what distance D are the two forces equal? (*Note.* The constant k is the same in each expression. The large numbers are the masses in kilograms of the earth and of the moon.)

8. Jessica is three years older than Benjamin. The product of their birth years is 3 863 188 years^2. How old is each? How could you have done this problem without knowing the difference between my children's ages?

*9. Two drivers play a game of chicken by driving toward one another and waiting to apply the brakes until the last instant. Suppose each car requires 180 feet to stop when it is traveling 60 mi/hr. The drivers hit the brakes simultaneously when the cars are 300 feet apart, and one car is traveling twice as fast as the other. The cars stop 10 feet apart. How fast was each going? What would have happened had each been driving 5 mi/hr faster?

10. A farmer has 120 feet of fencing and wishes to enclose a rectangular garden. If he makes it 20 feet long it will have to be 30 feet wide (to use all the fencing) and will thus have an area of 600 ft^2. But he can change the dimensions if he wishes.

(a) If the garden is L feet long, how wide is it?

(b) Find a formula showing how the area of the garden depends on its length L.

(c) Find the value of L that produces the garden of largest area. How large is this garden?

(d) What value of L produces the smallest garden?

(e) How large a garden will he have if he makes it circular?

***11.** Solve each of the following equations for P. If there are no solutions, say so. When the answers are numbers, check them. Notice that sometimes your answers will necessarily be expressed in terms of some constants, hence will not be numerical.

(a) $P^2 + 3P + 2 = 0$ (b) $P^2 - 3P + 2 = 0$

(c) $P^2 + 3P - 2 = 0$ (d) $P^2 - 3P - 2 = 0$

(e) $P^2 + 3P = 2$ (f) $P^2 + 2P - 3 = 0$

(g) $P^2 + 2P + 3 = 0$ (h) $P^2 + 4P + 4 = 0$

(i) $2P^2 + 3P + 6 = 0$ (j) $P^2 + 3P + c = 0$

(k) $P^2 + bP + C = 0$ (l) $aP^2 + bP + c = 0$

12. Suppose that $x + y = S$, and $xy = P$. (S stands for sum and P for product.)

(a) Find x and y if $S = 18$ and $P = 156$.

(b) Find x and y if $S = P = 4$.

(c) Find x and y in terms of S and P. Show that there is a solution only when $S^2 - 4P \geqslant 0$.

13. The *geometric mean G* of two positive numbers is defined as the square root of their product. Their *arithmetic mean A* is half their sum. Thus the arithmetic mean is just what we usually call the average.

(a) Find the arithmetic and the geometric means of 50 and 100.

(b) Find the arithmetic and the geometric means of 2.92 and 3.21.

(c) Find the arithmetic and the geometric means of 12.068 and 12.068.

(d) If the arithmetic mean of two numbers is 21 and their geometric mean is 17, what are the numbers? (You may want to use the techniques of Problem 12.)

(e) Use the result of Problem 12(c) to show that the geometric mean is always less than or equal to the arithmetic mean. When are they equal?

(f) Show that the arithmetic mean of $\log(x)$ and $\log(y)$ is the logarithm of the geometric mean of x and y.

***14.** (a) Figure 17.7 shows a rectangle from which a square has been cut. If the height of the rectangle is 1, how long should the

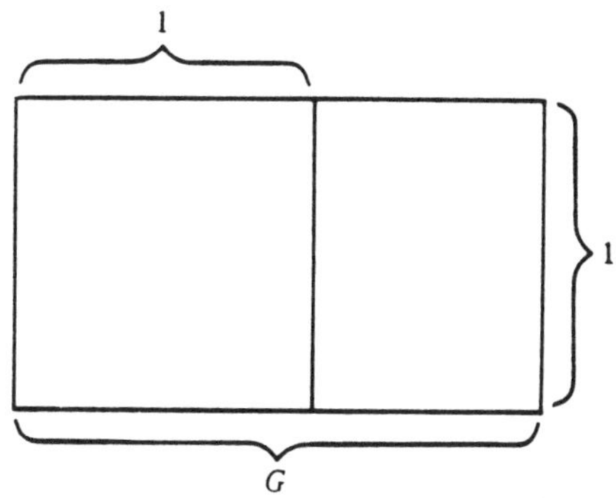

Figure 17.7 A golden rectangle

base G be in order to make the resulting small rectangle the same shape as the original from which it was cut? (*Hint*: Use Figure 17.7 to show that G must satisfy the equation:

$$\frac{G}{1} = \frac{1}{G-1}.$$

Then convert this into a quadratic equation and solve it for G.)

The Greeks called the answer G to this problem the *golden mean*. Rectangles like those in Figure 17.7, whose sides are in the ratio $G:1$, are said to be the most pleasing.

(b) How close does a standard 3 × 5 inch file card come to being the ideal shape?

(c) If we wanted to change our common 8.5 × 11 inch paper to a pleasing size while keeping the same height, how should we alter its width?

15. Use the quadratic formula to show that the peak of a parabola is always symmetrically located with respect to its roots.

Part IV
Trigonometry

CHAPTER 18

Right Triangles

I am but mad north-north-west; when the wind is southerly, I know a hawk from a handsaw.

– *Hamlet*, Act II, Scene 2

Surveying / Navigation / Propped ladders
Pythagoras remembered

EXAMPLE 1. THE WIDTH OF A RIVER

A surveyor stands on one bank of a river, directly opposite a tree on the far bank. He then moves 35 meters along the shore. When he looks back at the tree, his line of sight makes an angle of 51° with the river bank. How wide is the river?

Figure 18.1(a) shows a sketch of the scene and Figure 18.1(b) abstracts the relevant data from the sketch. It shows a right triangle ABT (with right angle at B). Angle A is 51°; side AB is 35 m. We must find the length of BT. To do this we first consider *all* the right triangles with a 51° angle. Some of them are drawn in Figure 18.2. This figure reminds us of direct proportion. Since the triangles are similar, the vertical length (the side *opposite* the 51° angle) is *proportional* to the horizontal length (the side *adjacent* to the 51° angle). Thus there is a constant k such that

$$\text{opposite side} = k * \text{adjacent side} \tag{18.1}$$

in any right triangle with a 51° angle. If we knew the numerical value of the proportionality constant k we could solve the surveyor's problem.

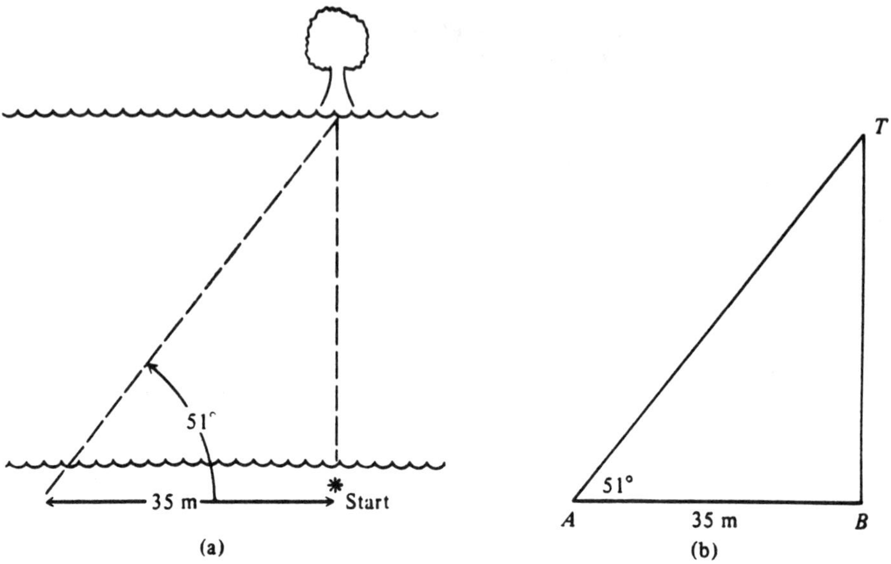

Figure 18.1 A wide river

If we made the same argument for some other angle, Equation 18.1 would still be true, but the value of k would be different. That is, in Equation 18.1 the proportionality constant k depends on the angle A. Fortunately, the dependence of k on A has been well studied. The constant k has a name: it is called the *tangent* of A, and written

$$\tan A \quad \text{or} \quad \tan(A).$$

Since we are interested in the case in which $A = 51°$, the number we need is

$$\tan 51°.$$

Your calculator knows how to find tangents. When you enter

51 [TAN]

the number

1.234897

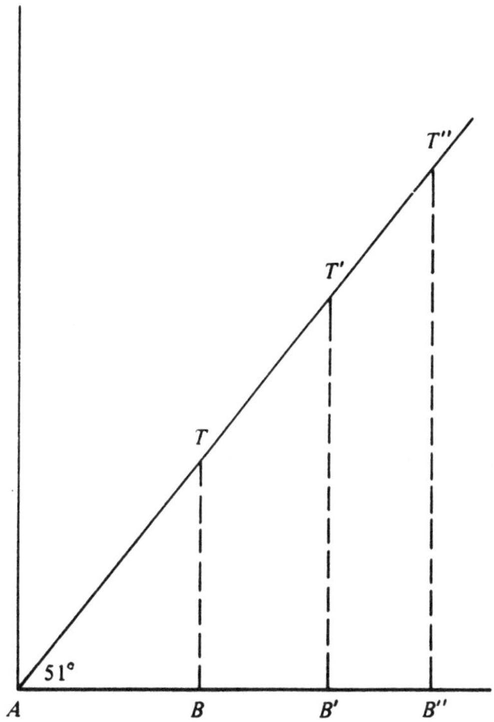

Figure 18.2 Some similar triangles

should appear in the display. Thus,

$$\tan 51 = 1.234897\ldots.$$

we read this as "tan 51 equals . . ."; to find it we enter the 51 first, then use the [TAN] key. Recall that logarithms are calculated in the same way.

Finally, substituting in Equation 18.1, we get

$$\begin{aligned} \text{width of river} &= \text{side opposite } 51^\circ \text{ angle} \\ &= \tan(51) * \text{side adjacent} \\ &= 1.2 * 35 \\ &= 43 \text{ meters.} \end{aligned} \qquad (18.2)$$

What we know about tangents can be summarized by rewriting

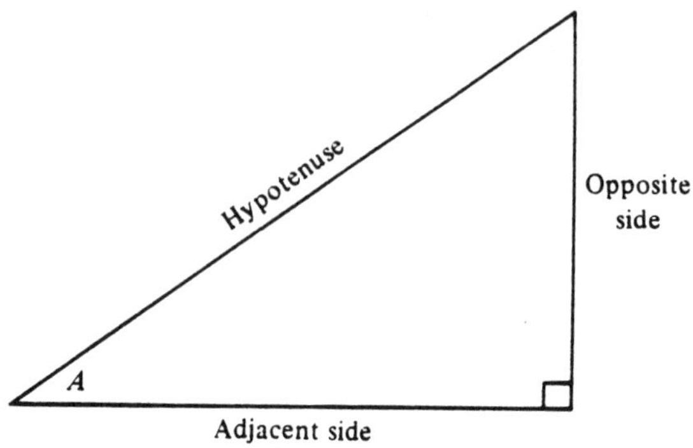

Figure 18.3 A right triangle

Equation 18.1. In a right triangle (Figure 18.3) in which A is one of the two acute angles,

$$\text{side opp. } A = \tan A * \text{side adj. to } A \tag{18.3}$$

or

$$\tan A = \frac{\text{opposite side}}{\text{adjacent side}}. \tag{18.4}$$

Since tan A is a ratio of lengths it has no units. Equation 18.4 tells us that tan A is the *slope* of a line that meets the horizontal axis at angle A.

Some useful facts about tangents follow from Equation 18.4. When A is a very small angle, the opposite side is small relative to the adjacent side, and tan A is small. When $A = 45°$, the opposite and adjacent sides are equal, so that

$$\tan 45 = 1.$$

As A increases toward 90°, tan A increases rapidly. These facts are illustrated by the triangles in Figure 18.4, which also shows that when A is negative, tan A is negative. Table 18.1 shows how tan A depends on the size of angle A. This dependence is also shown by the curve in Figure 18.5.

Here is a second surveying problem we can solve by using the tangent function. Suppose we are on one edge of a ravine and a 30-foot flagpole stands opposite us on the other edge. When we look up from ground level at the flag our line of sight makes an angle of 28° with a horizontal line. How wide is the ravine?

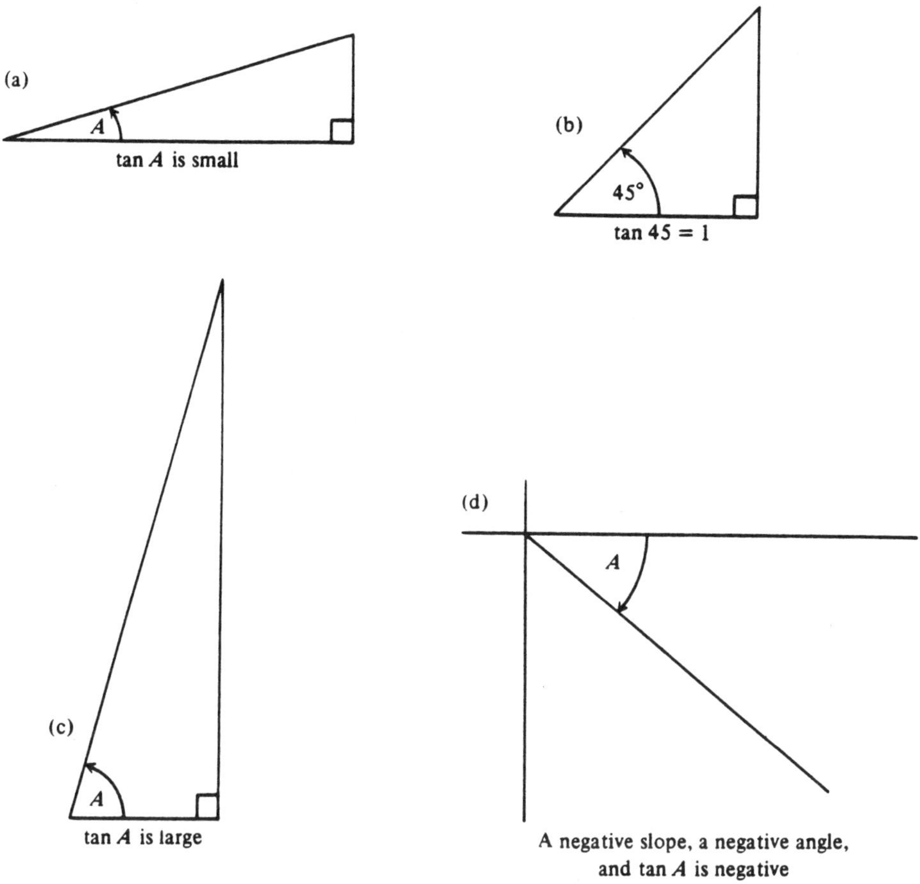

Figure 18.4 Angles of various sizes

This problem is pictured in Figure 18.6(a); Figure 18.6(b) displays the implicit right triangle. Here, the side opposite the 28° angle is 30 feet; the adjacent side R is unknown. This problem differs from the one in Figure 18.1 in that there the adjacent side was known and the opposite side unknown. We may still use Equation 18.4 (or Equation 18.3):

$$\tan 28 = \frac{30}{R}$$

so

$$R = \frac{30}{\tan 28} = 56 \text{ ft.}$$

Table 18.1 The tangent function

$A°$	tan A
0	0.00
30	0.58
45	1.00
60	1.73
75	3.73
80	5.67
85	11.43
89	57.29

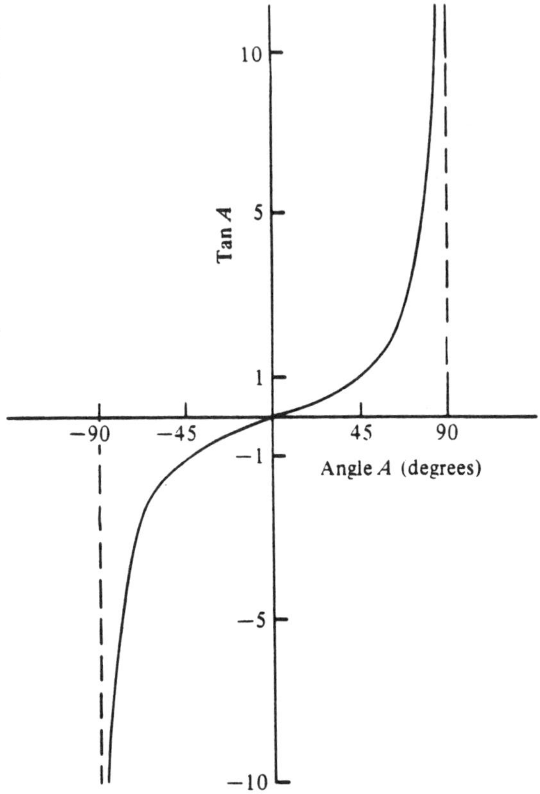

Figure 18.5 The tangent

EXAMPLE 2. NAVIGATION

A bushwhacking pilot is 133 miles west and 85 miles south of the airport. What course should he follow to reach it? How far away is he?

As in all trigonometry problems, we start with a diagram. Figure 18.7 shows the relative positions of the pilot and the airport. The unknowns in the problem are the distance D and the angle C, which is the desired compass course. The diagram tells us that we are interested in an angle and the sides opposite and adjacent to it, and so we will use the tangent function. Substituting in Equation 18.4 gives:

$$\tan C = \frac{133}{85} = 1.56. \tag{18.5}$$

What is C? This is a new kind of problem. In Example 1 we were given the angle and one of two sides, and asked for the other side. Here we are given the two sides and seek the angle. We must solve Equation 18.5 for C. The graph in Figure 18.5 shows that the answer

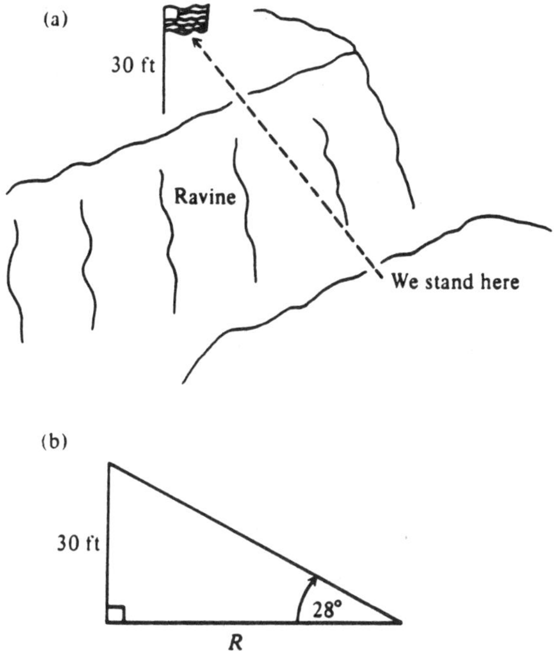

Figure 18.6 A wide ravine

is approximately 60°. To find it more exactly, we follow a procedure analogous to the one we used to define logarithms. We give the solution a *name*, and then rely on the fact that our calculators have been programmed to find its numerical value, by a sequence of trial and error approximations.

The solution to Equation 18.5 is the angle whose tangent is equal to 1.56; it is called

$$\text{Arctan } 1.56.$$

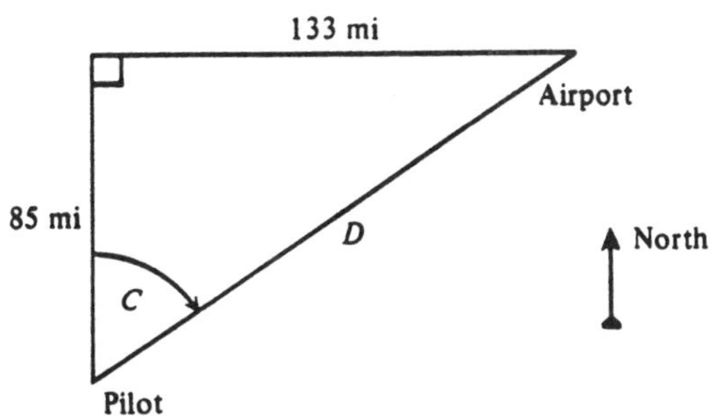

Figure 18.7 Navigation

To find the numerical value of Arctan 1.56, look on your calculator for a key labeled [ARCTAN] or [TAN^{-1}]. Recall the quotient 133/85 = 1.5647 . . . from the calculator's memory and then use the [ARCTAN] key:

1.5647 . . . [ARCTAN].

The display will show

57.4174 . . . ,

so that the pilot's course should be about 57°.

If your calculator has no [ARCTAN] or [TAN^{-1}] key, try

1.56 [INV] [TAN]

or

1.56 [2ND] [TAN].

In general, to solve the equation

$$\tan C = Q \qquad (18.6)$$

for C in terms of Q, just compute

$$C = \text{Arctan } Q. \qquad (18.7)$$

The relationship between the operations tan and Arctan is just like that between log and 10^x, or between squaring and taking square roots. Each inverts or undoes the operation of the other.

The last part of the problem asks for the distance to the airport. This is the length D of the hypotenuse of the right triangle in Figure 18.7. The Pythagorean theorem tells us that:

$$D^2 = 133^2 + 85^2 = 24\,914$$

so

$$D = \sqrt{24\,914} = 159 \text{ miles.}$$

EXAMPLE 3. THE LADDER

A 6.5-m ladder makes an angle of 62° with the ground. How high up the wall does it reach?

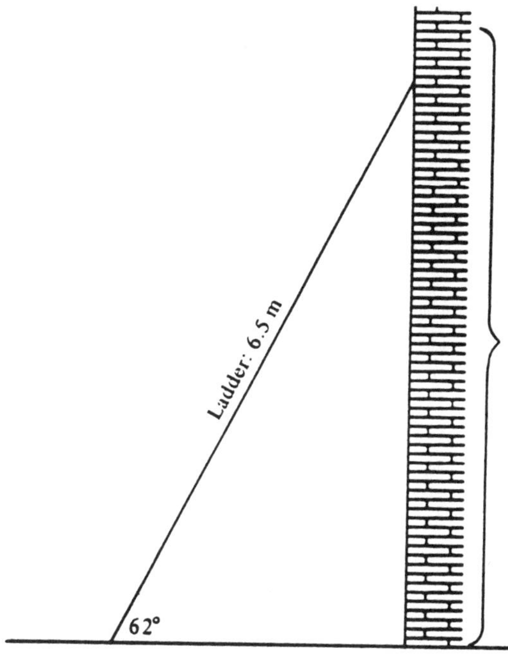

Figure 18.8 A ladder against a wall

As usual, we begin this trigonometry problem by drawing a diagram. Figure 18.8 shows the 6.5-m ladder, the 62° angle, and the unknown height H reached by the ladder. This problem resembles the one we solved in Example 1. There too we knew one side and one acute angle in a right triangle, and sought the length of the opposite side. Here, however, we are given the length of the hypotenuse rather than the length of the adjacent side. We shall consider all the right triangles with a 62° angle. These triangles, shown in Figure 18.9, are all similar, so there is a constant k such that

$$\text{opposite side} = k * \text{hypotenuse}. \tag{18.8}$$

The proportionality constant k is called the *sine* of 62°, and is written

$$\sin 62 \quad \text{or} \quad \sin(62).$$

The calculator finds sines with the [SIN] key. Enter

62 [SIN]

and see the answer

0.882948.

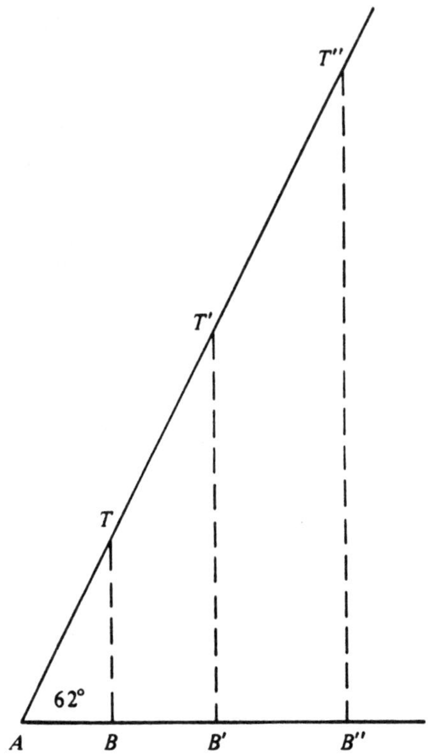

Figure 18.9 More similar triangles

Thus for the triangle we are interested in we know that

$$\begin{aligned} \text{height on wall} &= \text{side opposite } 62^\circ \text{ angle} \\ &= \sin(62) * \text{hypotenuse} \\ &= 0.88 * 6.5 \\ &= 5.7 \text{ m.} \end{aligned} \tag{18.9}$$

For an arbitrary right triangle in which A is one of the two acute angles,

$$\text{side opposite } A = \sin(A) * \text{hypotenuse}, \tag{18.10}$$

or

$$\sin A = \frac{\text{opposite side}}{\text{hypotenuse}}. \tag{18.11}$$

Table 18.2 shows some sample calculations of sines, which we use to draw the solid curve in Figure 18.10 showing how sin(A) depends on A. When angle A is close to 0°, the opposite side is very small relative to the hypotenuse, and sin(A) is small. When angle A is close to 90° the opposite side is nearly as large as the hypotenuse, and sin(A) is almost 1. Sin(A) can never be greater than 1, because the hypotenuse must be the longest side in any right triangle.

Suppose we prop our ladder so that its top is 4.1 m from the ground, and wish to compute the angle A?

Then in Equation 18.11 we have

$$\sin A = \frac{4.1}{6.5} = 0.63. \tag{18.12}$$

To solve this for A we must invert or undo the operation *sin*. The function that does this is the Arcsin. That is, if

$$\sin A = Q$$

then

$$A = \text{Arcsin } Q.$$

Thus the solution to Equation 18.12 is

$$A = \text{Arcsin } 0.63.$$

The solid curve in Figure 18.10 shows that A is about 40°.

Table 18.2 Sines and cosines

A°	sin A	cos A
0	0	1.000
5	0.087	0.996
30	0.5	0.866
45	0.707	0.707
60	0.866	0.5
85	0.996	0.087
90	1.000	0

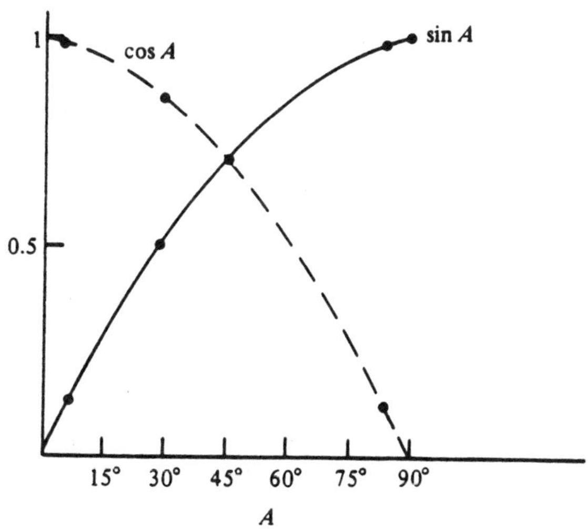

Figure 18.10 Graphs of sine and cosine

To find it more exactly, use the calculator key labeled [ARCSIN] or [SIN^{-1}], or one of the pairs [INV] [SIN] or [2ND] [SIN]:

0.63 [ARCSIN]

gives

$$39.11 \sim 39°.$$

Thus the ladder makes an angle of 39° with the ground. This is not a very safe angle. To reach 4.1 m a shorter ladder would be safer.

If you try to calculate Arcsin Q when Q is larger than 1, the calculator sends you an error message. This is because there is no angle whose sine exceeds 1. Finding Arcsin(2) is as impossible as finding $\sqrt{-3}$ or log(−4)

There is one more ratio that is useful in studying right triangles. If in Example 3, when angle A is 62°, we wish to figure out how far the foot of the ladder is from the base of the wall, we need to know the constant k for which

$$\text{side adjacent to } A = k * \text{hypotenuse}. \qquad (18.13)$$

This constant is called the *cosine* of A, and is written

$$\cos A \quad \text{or} \quad \cos(A).$$

Use the [COS] key on your calculator to find its value. Thus,

$$\begin{aligned} \text{distance from ladder to wall} &= \text{side adjacent to } 62° \text{ angle} \\ &= \cos(62) * \text{hypotenuse} \\ &= 0.47 * 6.5 \\ &= 3.1 \text{ m}. \end{aligned} \qquad (18.14)$$

In general,

$$\text{side adjacent to } A = \cos(A) * \text{hypotenuse}, \qquad (18.15)$$

or

$$\cos A = \frac{\text{adjacent side}}{\text{hypotenuse}}. \tag{18.16}$$

Column 3 of Table 18.2 and the dashed curve in Figure 18.10 show how cos(A) depends on A.

EXAMPLE 4. RELATIONS AMONG SIN, COS, AND TAN

The sine, cosine, and tangent of an angle A are the three possible ratios between pairs of sides of a right triangle in which one acute angle is equal to A. Each is useful in a different kind of problem. In this example we explore briefly how, if you knew just one of the three, you could find the others. Since all right triangles with a given angle A are similar, we can learn all we need to know by studying one of them—the one whose hypotenuse has length 1. (This could be 1 inch, or 1 cm, or 1 of any other unit of length.) In such a triangle, we have

$$\begin{aligned} \sin A &= \frac{\text{side opposite } A}{\text{hypotenuse}} \\ &= \text{side opposite } A, \end{aligned} \tag{18.17}$$

and

$$\begin{aligned} \cos A &= \frac{\text{side adjacent to } A}{\text{hypotenuse}} \\ &= \text{side adjacent to } A. \end{aligned} \tag{18.18}$$

Thus the labels in Figure 18.11 are correct.

Let us see what the Pythagorean theorem tells us about sin A and cos A. In Figure 18.11 the sum of the squares of the legs is the square of the hypotenuse, so

$$(\sin A)^2 + (\cos A)^2 = 1^2 = 1. \tag{18.19}$$

Thus, if we knew sin A, we could find cos A by calculating

$$\cos A = \sqrt{1 - (\sin A)^2}. \tag{18.20}$$

Equation 18.20 shows again that sin A must be less than 1. Otherwise

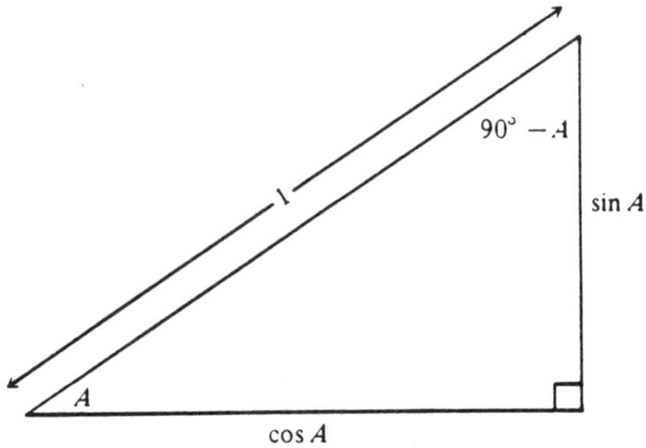

Figure 18.11 A model right triangle

the equation would force us to find the square root of a negative quantity.

Before we look at tan A, think for a moment about the third angle of the triangle in Figure 18.11. Since the sum of the angles of a triangle is 180°,

$$A + B + 90° = 180°$$

and hence

$$B = 90° - A.$$

Thus, A and B are complementary angles. But the side adjacent to A is the one opposite B and vice versa, so

$$\sin B = \cos(90° - B) = \cos A$$

and

$$\cos B = \sin(90° - B) = \sin A.$$

This explains why the third column of Table 18.2 is just the second column read from bottom to top.

Once we know sin(A) and cos(A), it is easy to find tan(A). Look again at Figure 18.11. By definition,

$$\tan A = \frac{\text{opposite side}}{\text{adjacent side}},$$

so

$$\tan A = \frac{\sin A}{\cos A}. \tag{18.21}$$

Let us look at the values of sin(A), cos(A), and tan(A) for some specific values of angle A. When A happens to be 45°, the triangle in Figure 18.11 is isosceles. Then tan(45) = 1 (which we already knew), because

$$\sin 45 = \cos 45.$$

Substituting in Equation 18.19 gives

$$(\sin 45)^2 + (\sin 45)^2 = 1,$$

so

$$2(\sin 45)^2 = 1,$$

and

$$\sin 45 = \sqrt{1/2} = 0.707. \tag{18.22}$$

When angle $A = 60°$, we can compute sin(60), cos(60), and tan(60), by studying the equilateral triangle shown in Figure 18.12. We draw this

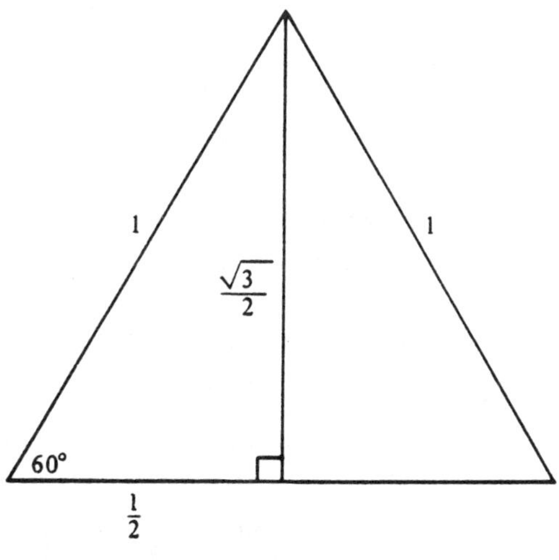

Figure 18.12 An equilateral triangle

triangle so that its side has length 1. Then the left half of the triangle is a right triangle with a hypotenuse of 1 and with a side of 1/2 adjacent to the 60° angle. Therefore,

$$\cos 60 = \frac{\text{adjacent side}}{\text{hypotenuse}} = \frac{1}{2}.$$

Then

$$\sin 60 = \sqrt{1 - (\cos 60)^2} = \sqrt{\frac{3}{4}} = \frac{\sqrt{3}}{2},$$

and

$$\tan 60 = \frac{\sin 60}{\cos 60} = \sqrt{3}.$$

PROBLEMS

*1. Hamlet sights a hawk hovering over Bald Mountain four miles away. If he is looking up at an angle of 15°, how high is the hawk?

2. When the shadow of a 60-foot fir tree is 37 feet long, what angle do the sun's rays make with the ground? Answer the same question when there is no shadow.

*3. A kite is aloft on a 300-meter string that makes an angle of 52° with the ground. How high is the kite? If a piece of its tail drops off and falls straight to the ground, how far from the kite flyer will it land?

4. Suppose you know the value of a quantity Q.
   (a) What would you name the solution A to the equation $\cos A = Q$?
   (b) How would you find it with your calculator?
   (c) What range of values can Q have?
   (d) If $\cos(A) = 0.75$, what is the value of A? What if $\cos(A) = 1.23$?

5. A 130-foot radio tower is to be anchored by cables that should make an angle of 50° with the ground. How long must the cables be? How far from the foot of the tower should the anchors be set?

*6. Looking out over Penobscot Bay from the lighthouse, Hog Island is 4 miles away at an angle 23° west of south. How far south of the lighthouse is the island?

7. Blue Hill is 500 ft high. When I look at it from my window, I look up at an angle of 30°. How far away is Blue Hill?

8. The moon is 240 000 miles away. When you look at it in the night sky the light rays from opposite edges make an angle of half a degree at your eye.

   (a) Let D be the diameter of the moon. Draw a diagram with which you can justify the approximation

   $$\sin 0.5 \sim \frac{D}{240\,000}.$$

   Then use the approximation to find D.

   (b) Use the fact that the sun and the moon look to us to be about the same size, and the distance to the sun (given in Chapter 8), to calculate the diameter of the sun.

   (c) The great nebula in Andromeda is 2 million light-years away and subtends an angle of 3° in the sky. (This is 6 times the angle subtended by the sun or the moon.) What is its diameter?

9. The base of a 16-foot ladder is 4 feet 9 inches from the wall against which it rests. How high on the wall does it reach? What angle does it make with the wall? With the ground?

*10. Two boats leave Honolulu at noon. One steams south at 20 knots. The other sails east at 4 knots.

   (a) What is a knot?

   (b) How far apart are the ships at 5 P.M.?

   (c) What is the compass course from the steamship to the sailboat at 5 P.M.? From the sailboat to the steamship?

11. Verify the identity:

   $$(\sin A)^2 + (\cos A)^2 = 1$$

   for the angles A whose sines and cosines are given in Table 18.2.

12. It is 90 feet from home plate to first base. How far is it from home plate to the pitcher's mound? (See Problem 4 of Chapter 17.)

13. Verify the identity:

$$1 + (\tan A)^2 = \frac{1}{(\cos A)^2}.$$

14. Is $(\sin A)^2$ the same as $\sin(A^2)$? Is $2 \sin(A)$ the same as $\sin(2A)$? Do some sample calculations to find out. If you go on in mathematics you will learn that $(\sin A)^2$ is usually written as $\sin^2 A$ in order to avoid confusing it with $\sin(A^2)$.

*15. Figure out how to find $\tan(90 - A)$ in terms of $\tan(A)$.

16. Check that for very small angles $\sin A \sim \tan A$. Why should this be?

17. Suppose that $\sin(A) = 0.32$. Find $\cos(A)$ and $\tan(A)$ without first finding A, and without using any of the trigonometric or inverse trigonometric keys on your calculator.

18. Look up the etymology of the word *trigonometry*.

19. One can use cosines even when the triangles to be studied are not right triangles. In the triangle in Figure 18.13 it is true that

$$c^2 = a^2 + b^2 - 2ab(\cos C). \tag{18.23}$$

This fact is known as the *law of cosines*. We shall make use of it, but not take the time to learn why it is true.

(a) If angle C should happen to be a right angle, show that the law of cosines becomes the Pythagorean theorem.

(b) Suppose that side $a = 45$, $b = 52$, and angle $C = 66°$. Find the third side of the triangle.

*20. (a) Find all three angles in a triangle whose sides are 3, 4, and 5.

(b) Find all three angles in a triangle whose sides are 4, 5, and 6.

21. Two boats leave Honolulu at noon. One steams on a course 6° east of south at 20 knots. The other sails 6° south of east at 4 knots.

(a) How far apart are the ships at 5 P.M.?

(b) What is the compass course from the steamship to the sailboat at 5 P.M.? From the sailboat to the steamship?

*22. Hamlet sights a hawk to the north-north-west four miles away. The wind comes from Elsinore, six miles to the south of him. How far is the hawk from Elsinore? (*Hint*: The angle at Hamlet in the Hawk-Hamlet-Elsinore triangle is 157.5°. (Why?) Notice that the law of cosines works even in triangles where one angle exceeds 90°.)

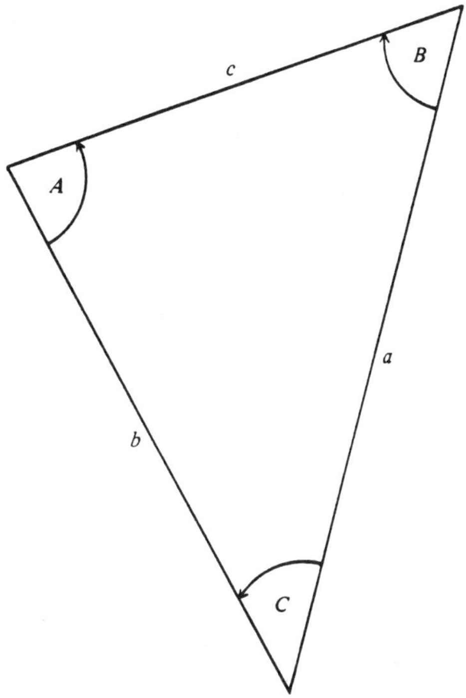

Figure 18.13 A typical triangle

23. One can use sines even when the triangles to be studied are not right triangles. In the triangle in Figure 18.13 it is true that

$$\frac{\sin A}{a} = \frac{\sin B}{b} = \frac{\sin C}{c}. \tag{18.24}$$

This fact is known as the *law of sines*. We shall make use of it, but not take the time to learn why it is true.

(a) If angle C should happen to be a right angle, show that the law of sines tells you something you already know.

(b) Suppose angle A is 37°, angle B is 56°, and the side c between them is 8.3 feet long. Find the third angle and the other two sides of the triangle.

***24.** The surveying class has laid out a 200-yard baseline, from the drumlin to the barn. The angle between this baseline and the line from the barn to the flagpole is 71°; the angle between the baseline and the line from the drumlin to the flagpole is 57°. How far is the flagpole from the barn? From the drumlin?

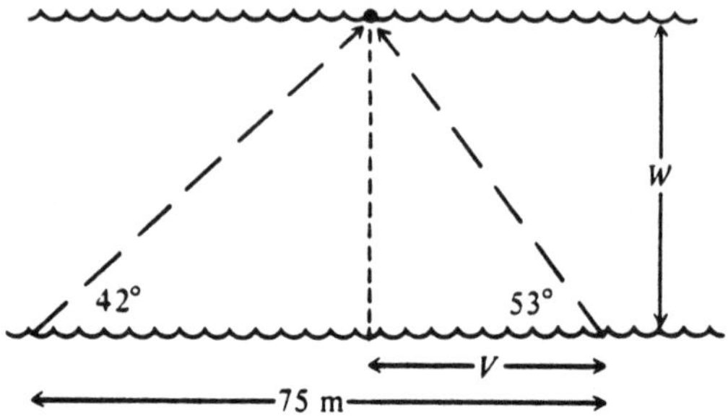

Figure 18.14 A surveyor's sketch

25. Figure 18.14 shows the results of a surveyor's measurements. Find W, the width of the river. (*Hint*: You may want to find V first, even though it is not called for.)

CHAPTER 19

Oscillations

Up and down, up and down,
I will lead them up and down;
I am fear'd in field and town.
Goblin, lead them up and down.
– *A Midsummer Night's Dream*, Act III, Scene 2

Rotation, waves, and springs / Radian measure
Predators and prey

EXAMPLE 1. ROTATION

Imagine a vertically mounted circular tape reel with a radius of 1 foot. As it spins counterclockwise a spot on the rim travels at a speed of 1 ft/sec. Study the horizontal and vertical motions of the spot.

We can see that the spot moves up and down and back and forth simultaneously. To understand this motion, let us measure the horizontal and vertical position at various times. We shall use the center of the reel as a reference point, and suppose that at time $t = 0$ the spot happens to be at the right edge of the reel, as shown in Figure 19.1. At that time the horizontal displacement H is 1 ft; the vertical displacement V is 0 ft. What are the values of H and V after 1.2 sec? Then the spot has moved 1.2 ft on the rim, and occupies the position shown in Figure 19.1. Then H and V are the two legs of the triangle shown. Since the hypotenuse is 1 foot, if we knew the angle A, we could calculate the legs:

$$H = \sin A \quad \text{and} \quad V = \cos A.$$

To find the angle A, we use the fact that the reel turns at a constant rate. Thus the size of A, measured in degrees, is proportional to the distance D the spot has traveled on the rim:

$$A = kD. \tag{19.1}$$

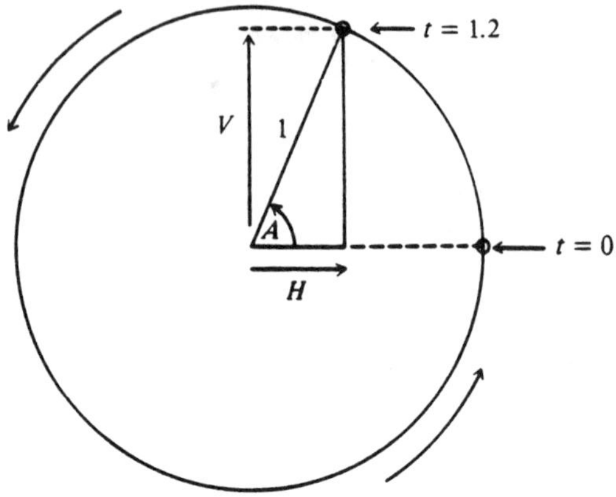

Figure 19.1 The tape reel

We need only know the proportionality constant k. To find it, first calculate the circumference of the reel. This is 2π times its radius, or

$$2\pi * 1 = 2\pi \sim 6.28 \text{ ft.} \tag{19.2}$$

Thus in about 6.3 seconds the spot returns to its starting point. At that time it will have traveled a full 360°. Therefore $A = 360°$ when $D = 2\pi$. This is enough information for us to find k:

$$360° = k * 2\pi,$$

so

$$k = \frac{360°}{2\pi} = 57.295780 \sim 57°. \tag{19.3}$$

Therefore,

$$A = 57D. \tag{19.4}$$

Thus for any distance D measured along the rim of the reel,

$$H = \sin(57D) \quad \text{and} \quad V = \cos(57D).$$

In particular, when $D = 1.2$,

$$H = \sin(57 * 1.2) = 0.93, \tag{19.5}$$

and

$$V = \cos(57 * 1.2) = 0.36.$$

If we have many such calculations to make, it is easier to tell the calculator to make the conversion from the value of the distance D along the rim to the angle A (in degrees) automatically.

> The calculator feature that does this job is a key (or switch) labeled [RAD] or [RADIANS]. If you first tell your calculator you are using radian measure and then enter
>
> 1.2 [SIN]
>
> the result will be
>
> 0.932039,
>
> the answer we found in Equation 19.5.

We have just changed the unit we use to measure angles, from the customary degree to a new unit called the *radian*. Equation 19.4 tells us that

$$1 \text{ radian} = 57.295780° \sim 57.3°.$$

A complete circle is

$$360° \quad \text{or} \quad 2\pi \text{ radians},$$

while a right angle, which is a quarter of a circle, measures

$$90° \text{ or } \frac{2\pi}{4} = \frac{\pi}{2} \text{ radians}.$$

Since a 45° angle is half of a right angle, it measures $\pi/4$ radians. The angle in an equilateral triangle is 60°, which is 1/6 of a circle, as shown in Figure 19.2, so

$$60° = \left(\frac{1}{6}\right)2\pi = \frac{\pi}{3} \text{ radians}.$$

This figure will also help you remember why 1 radian is a little less than 60°. If you travel 1 foot in a straight line from A you reach B. If you travel along the rim you reach only C, one radian away.

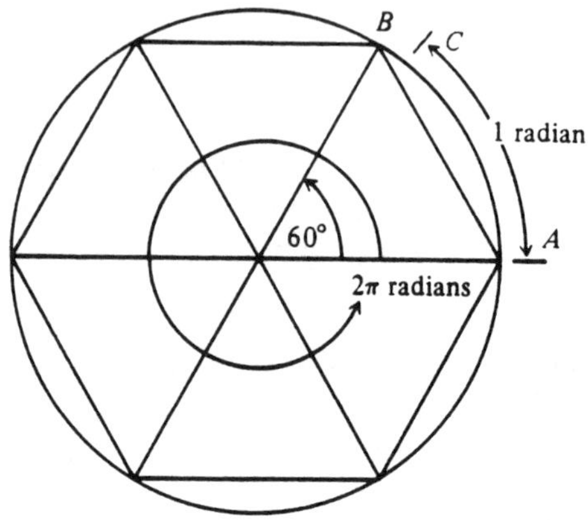

Figure 19.2 Equilateral triangles in a circle

When we were studying triangles in Chapter 18 it was natural to measure angles in degrees. When we study rotation or oscillation we traditionally measure angles in radians, and tell our calculators that we are doing so. With our calculators so instructed, we can work out the data in Table 19.1.

Notice that when we measure angles in radians it is sensible and easy to consider angles greater than a full circle. They correspond to rotating the tape reel more than one full turn. Figure 19.3 graphs the data in Table 19.1. The two curves show H and V, alias sin(D) and cos(D), oscillating back and forth between the values +1 and −1, the farthest horizontal and vertical displacements from the center of the reel. Every

Table 19.1 Values of sines and cosines when measuring angles in radians

D (radians)	sin D	cos D	angle (degrees)
0	0.00	1.00	0
1	0.84	0.54	57.30
$\pi/3$	0.87	0.5	60
$\pi/2$	1.00	0.00	90
2	0.91	−0.42	114.59
3	0.14	−0.99	171.89
π	0.00	−1.00	180
4	−0.76	−0.65	229.18
$3\pi/2$	−1.00	0.00	270
5	−0.96	0.28	286.48
6	−0.28	0.96	343.77
2π	0.00	1.00	360
8	0.99	0.14	458.37
6π	0.00	1.00	1080
20	0.91	−0.42	1145.92

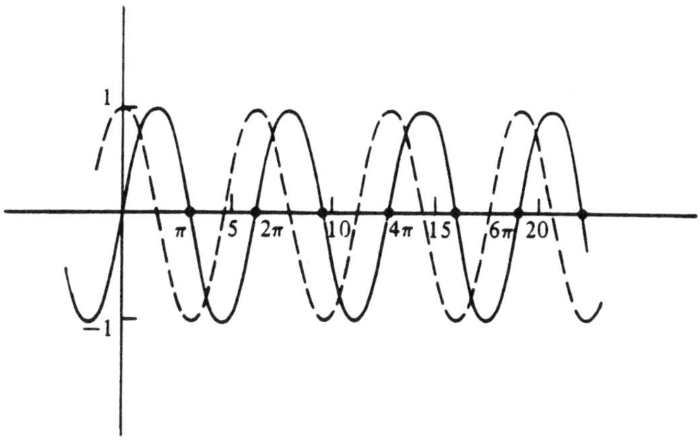

Figure 19.3 Graphs of sin (solid curve) and cosine (dashed curve)

2π seconds the pattern repeats, because every 2π seconds the spot on the reel returns to its starting place. The motion is *periodic*, with a *period* of 2π, or about 6.3 seconds.

EXAMPLE 2. SPRINGS

A weight on a spring bobs up and down. Describe its motion.

Let Q represent the distance of the weight above or below its natural resting place. When the spring is disturbed, Q changes with time t. If the weight is given an initial upward push of just the right strength at $t = 0$ seconds, the displacement Q will depend on t according to the formula:

$$Q = \sin(t), \tag{19.6}$$

where Q is measured in meters. The graph in Figure 19.4 shows that the weight has a maximum displacement from its rest position of 1 m, and that it returns to that rest position after π seconds. Then it is moving down. After another π seconds it returns again, moving up this time, and the whole cycle repeats itself. This kind of behavior is called *simple harmonic motion*. As in Example 1, the period of the motion is $2\pi = 6.28$ seconds.

Notice that this must be a very long, relaxed spring, since it takes more than 6 seconds to complete a 1-meter oscillation. I chose to discuss such an unrealistic spring to free us from the awkwardness of making unit conversions in Equation 19.6. Now we shall see how to

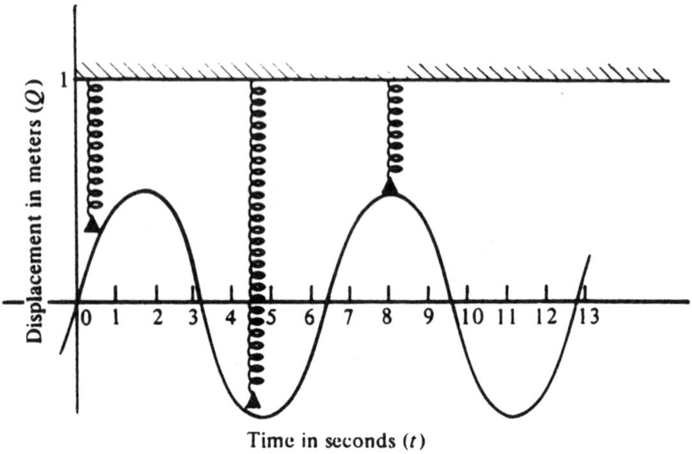

Figure 19.4 Graph of a bobbing spring

modify that equation so that it can describe springs of any size and stiffness.

Suppose we had pushed the spring twice as hard when we started it off. Then its motion would be described by the formula:

$$Q = 2\sin(t).$$

The solid curve in Figure 19.5 is the graph of the new motion. The bobs are twice as high, but the period has not changed. This figure also shows the graph of the motion

$$Q = A\sin(t) \tag{19.7}$$

for other values of the constant A, which is called the *amplitude* of the motion. The amplitude measures the height of the peaks. In this example the amplitude depends on how hard the spring is pushed at the start. Notice, though, that the strength of the push does not affect the period.

Suppose we can somehow stiffen the spring. Then when we start it bobbing its displacement may be given by the formula:

$$Q = \sin(1.3t). \tag{19.8}$$

In Figure 19.6 we compare the graph of this equation with the by now familiar sine curve. We see that the new curve has the same amplitude, 1, but bounces more quickly. It takes only 4.8 seconds to return to its starting point, moving up. This is a new, shorter period. You can cal-

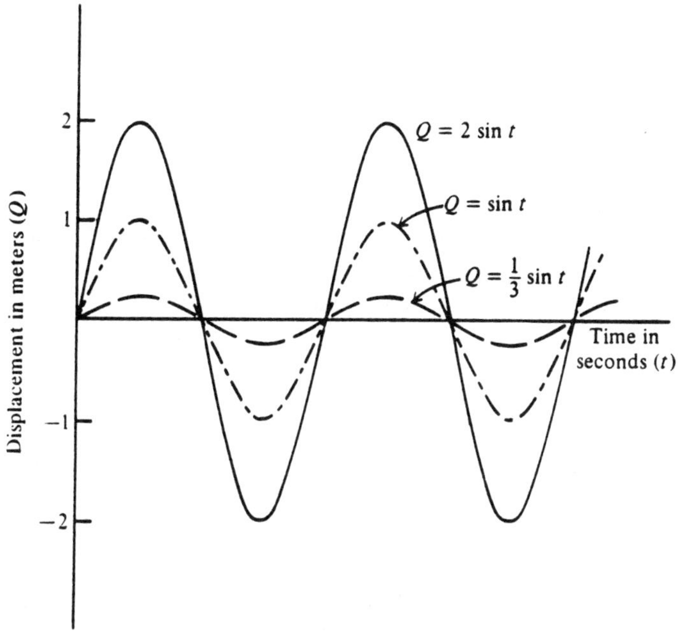

Figure 19.5 Miscellaneous amplitudes

culate the new period by solving for t the equation:

$$1.3t = 2\pi.$$

In general, to find the period L of the oscillation

$$Q = \sin(\omega t)$$

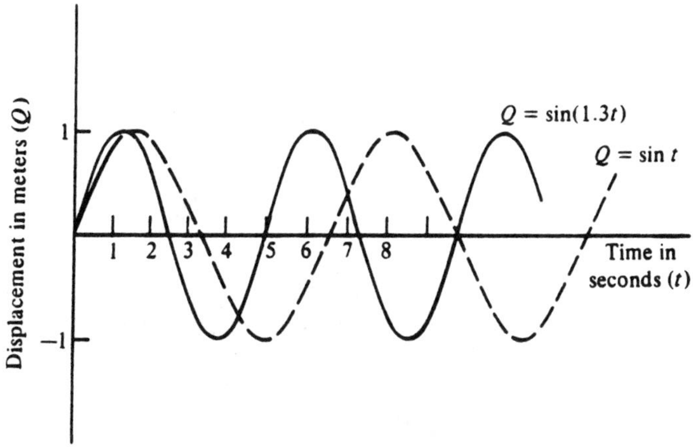

Figure 19.6 A stiffer spring

we solve

$$\omega L t = 2\pi$$

for L, giving

$$L = \frac{2\pi}{\omega}.$$

The symbol ω is *omega*, the last letter of the Greek alphabet. The number $\omega/2\pi$ is commonly called the *frequency* of the oscillation. For example, the oscillation whose displacement formula is

$$Q = 0.23 \sin(15t)$$

has an amplitude of 0.23 meters (or 23 cm), a period of $2\pi/15 = 0.42$ seconds, and a frequency of 1/0.42 or 2.4 cycles per second. It is illustrated in Figure 19.7, with the basic sine curve, dashed, for comparison.

We have one last way to generalize Equation 19.6. Suppose that, instead of giving the weight an upward push, we gently raise it one meter and then let go. It will then fall toward its rest position, continue beyond for one meter, and reverse direction to return to the point at which we let it go. The formula governing this motion will now be

$$Q = \sin\left(t + \frac{\pi}{2}\right) \sim \sin(t + 1.6).$$

The corresponding curve is drawn in Figure 19.8. It has the same shape as the basic sine curve, but it is shifted to the left by $\pi/2 = 1.6$ seconds. We recognize it as the graph of the cosine. In fact, one might shift the curve to the right by any amount ϕ by using the formula

$$Q = \sin(t - \phi). \tag{19.9}$$

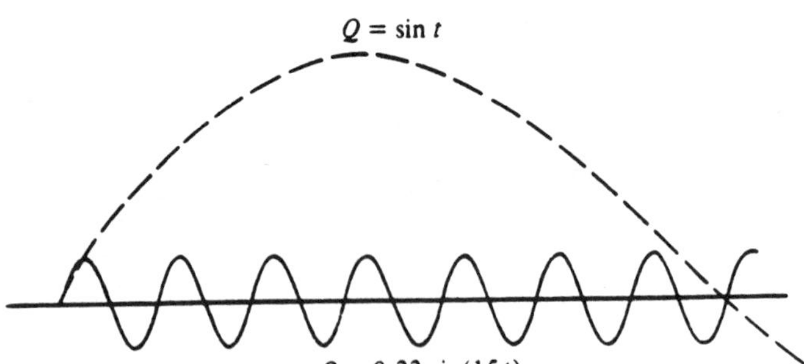

Figure 19.7 Fast oscillations

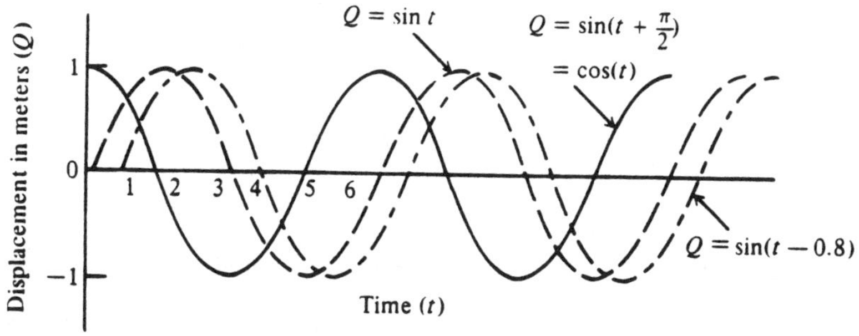

Figure 19.8 Out-of-phase sine curves

The symbol ϕ is the Greek letter phi. In this context it is referred to as the *phase*, or *phase angle*, of the oscillation. Figure 19.8 shows a sine curve with a phase of 0.7.

We can combine the three variations we have discussed so far in a single oscillation formula. For example, suppose

$$Q = 2 \sin[3(t - 1.5)]. \tag{19.10}$$

When we compare the graph of this relationship with the basic sine curve three facts emerge. The new graph is twice as high, so that its amplitude is 2. It oscillates 3 times as fast, so its period is $2\pi/3$, or 2.1 seconds. Finally, $Q = 0$ when $t = 1.5$, so the graph is out of step with that of the basic sine curve. The phase is 1.5 seconds. All this information shows in Figure 19.9.

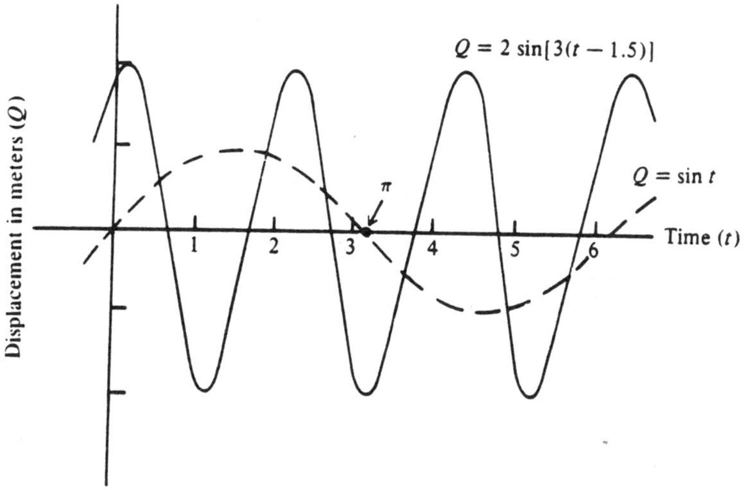

Figure 19.9 A typical oscillation

In general, by adjusting the amplitude A, the constant ω, which determines the frequency, and the phase ϕ, we can capture the motion of any oscillating spring or rotating wheel with the formula:

$$Q = A \sin[\omega(t - \phi)]. \tag{19.11}$$

EXAMPLE 3. HOW MUCH SUN IS THERE ON MY BIRTHDAY?

On February 23 in Boston the sun rises at 6:28 A.M. and sets at 5:28 P.M. How many hours of daylight are there on June 11?

To answer the question we need to know first that H, the number of hours of daylight, oscillates harmonically about its average value of 12. Thus H depends on D, the specific day of the year, according to the formula:

$$H = 12 + A \sin[\omega(D - \phi)]. \tag{19.12}$$

At the Vernal Equinox, March 21, there are 12 hours of daylight and the days are growing longer. This corresponds to the rest position of the spring (no pun intended) in the previous example. Since March 21 is the 80th day of the year the phase ϕ is 80.

The period of the oscillation is 1 year, or 365 days, so that

$$\frac{2\pi}{\omega} = 365,$$

and

$$\omega = \frac{2\pi}{365} = 0.0172.$$

Now we must find the amplitude A. We know that on February 24, when $D = 55$, there are 11 hours of daylight. Thus, substituting in Equation 19.12, we get

$$11 = 12 + A \sin[0.0172(55 - 80)].$$

Then

$$A \sin[0.0172(55 - 80)] = -1$$

so

$$A = \frac{-1}{\sin[0.0172(55 - 80)]}$$
$$= \frac{-1}{\sin(-0.43)}$$
$$= 2.4 \text{ hours.}$$

This amplitude of 2.4 hours means that in Boston the longest day has $12 + 2.4 = 14.4$ hours of sun, and the shortest has $12 - 2.4 = 9.6$ hours of sun.

On June 11, the 162nd day of the year, the number of hours of daylight is

$$H = 12 + 2.4 \sin[0.0172(162 - 80)]$$
$$= 14.37 \text{ hours}$$
$$= 14 \text{ hours } 22 \text{ minutes}$$

Notice that the number of hours of daylight is changing most rapidly at the equinoxes and least rapidly at the solstices.

PROBLEMS

*1. Convert the following angles, measured in degrees, to radians. When possible, express your answer as a simple multiple of π.

(a) $37.9°$ (b) $36°$
(c) $0°$ (d) $-123°$
(e) $135°$ (f) $180°$
(g) $79.83°$ (h) $120°$
(i) $57.3°$ (j) $-180°$

*2. Convert the following angles, measured in radians, to degrees.

(a) 1.5 (b) 2π
(c) -1.731 (d) $\frac{\pi}{8}$
(e) 0 (f) $\frac{2\pi}{3}$
(g) 90 (careful!) (h) $\frac{\pi}{6}$
(i) 1 (j) 1024π

*3. Find the amplitude, the period, the frequency, and the phase for each of these simple harmonic motions:

(a) $Q = 0.67 \sin(t)$
(b) $Q = \sin(0.67t)$
(c) $Q = \sin(t - 0.67)$
(d) $Q = 1.4 \sin[1.4(t - 1.4)]$
(e) $Q = 1.4 \sin(1.4t - 1.4)$
(f) $Q = \sin(\pi t)$
(g) $Q = \sin(t + 2)$
(h) $Q = \sin\left(t - \frac{\pi}{2}\right)$
(i) $Q = -\sin(t)$
(j) $Q = \alpha \sin[\beta(t - \gamma)]$

4. (a) Write the equation governing a simple harmonic oscillator with amplitude 5.4, period 3, and phase 1.6.
(b) Write the equation governing a simple harmonic oscillator with amplitude 5.4, period 1/3, and phase 1.6.
(c) Write the equation governing a simple harmonic oscillator with amplitude 5.4, period 3, and phase −1.6.
(d) Write the equation governing a simple harmonic oscillator with amplitude 5.4, frequency 3, and phase 1.6.

5. (a) Write a formula analogous to Equation 19.12 and use it to find the number of hours of daylight on your birthday. (Mimic the calculation in Example 3 for your home town. The phase and the frequency will be the same. All you have to do is figure out the amplitude.)
(b) What does Equation 19.12 look like for a town on the Arctic Circle? At the North Pole?

6. High tide in Boston on October 27, 1981 occurred at 10:44 A.M. and again at 11:08 P.M. Determine the period and the phase of the rising and falling of the tide. Can you find the amplitude from the data given? When did high tide occur at 5 A.M.? When did it next occur at about 11 P.M.?

7. In Chapter 14 we studied how populations grow. Now we can study how they fluctuate. When rabbits are plentiful, foxes multiply. That makes the rabbit population decrease, so the foxes go hungry and fail to reproduce. That in turn allows the rabbit population to grow once again. This phenomenon is captured in the following formulas for the rabbit and fox populations R and F:

$$R(t) = 100 + 10 \cos(1.5t),$$

while

$$F(t) = 12 + 2 \sin(1.5t).$$

Here R and F are measured in number of animals per acre, and the time t in years.

(a) Draw graphs of these curves on the same set of coordinate axes.

(b) What is the period of the population cycle?

(c) At what times is the rabbit population growing most rapidly? The fox population?

8. Any recurring phenomenon can be modeled by a combination of waves of the form given in Equation 19.11. For example, the curve in Figure 19.10 is the graph of

$$Q = \sin(t) + \tfrac{1}{2} \sin[2(t - 1)].$$

Carefully draw a graph of

$$Q = \sin(t) + \tfrac{1}{3} \sin(3t) + \tfrac{1}{5} \sin(5t) \qquad (19.13)$$

for values of t between 0 and 10 (radians). Your graph should end up approximating what electrical engineers call a *square wave.* What terms would you add to the right side of Equation 19.13 to improve the approximation?

9. In the 20 years from 1950 to 1970 the dollar cost C of a share of a conservatively growing stock varied according to the formula:

$$C(t) = 55 + 5t + 2 \sin(6t).$$

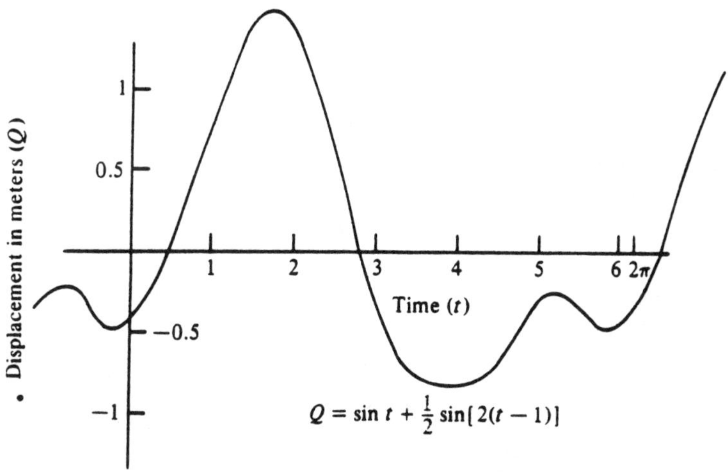

Figure 19.10 A complicated wave

(a) Draw a graph showing how C varied during the time in question.

(b) Describe the behavior you observe. What are the period and the amplitude of the oscillation? What curve is the value of C oscillating about?

(c) What would you expect the graph of this equation to look like:

$$I + bt + A \sin(\omega t)?$$

10. In the 20 years from 1950 to 1970 the dollar cost R of a share of a risky growth stock varied according to the formula:

$$R(t) = 55\ e^{0.12t} - 20 \sin(6t).$$

(a) Draw a graph showing how R varied during the time in question.

(b) Describe the behavior you observe. What are the period and the amplitude of the oscillation? What curve is the value of R oscillating about?

(c) What would you expect the graph of this equation to look like:

$$Ie^{rt} + A \sin(\omega t)?$$

11. Springs don't keep bobbing up and down forever. Their oscillations gently die out. Sketch the graph of the dependence:

$$Q(t) = 10e^{-0.05t} \sin(t)$$

for t between 0 and 20. Why do you think the result is called *damped oscillation*?

12. It is tempting to believe that all sorts of operations can be pushed through parentheses, as in

$$a * (x + y) = a * x + a * y.$$

Unfortunately,

$$\begin{aligned} (x+y)^2 &\text{ is not } x^2 + y^2; \\ \sqrt{x+y} &\text{ is not } \sqrt{x} + \sqrt{y}; \\ e^{(x+y)} &\text{ is not } e^x + e^y; \\ \log(x+y) &\text{ is not } \log x + \log y; \end{aligned}$$

and, of course,

$$\sin(x + y) \text{ is not } \sin x + \sin y.$$

However, there is a formula that tells what $\sin(x + y)$ is:

$$\sin(x + y) = \sin(x) * \cos(y) + \cos(x) * \sin(y) \qquad (19.14)$$

(a) Use your calculator to verify Equation 19.14 for several values of x and y.

(b) What interesting conclusion can you draw when you let $y = \frac{\pi}{2}$ in Equation 19.14? (See Figure 19.8.)

(c) What interesting conclusion can you draw when you let $y = \pi$ in Equation 19.14?

13. The addition formula for the cosine that corresponds to Equation 19.14 is:

$$\cos(x + y) = \cos(x) * \cos(y) - \sin(x) * \sin(y). \qquad (19.15)$$

The Swiss mathematician Leonhard Euler (1707–1783) discovered a truly remarkable connection between the sine and cosine functions and the number e. He found good reason to believe that if he let i stand for the square root of -1, so that

$$i^2 = -1,$$

then

$$e^{it} = \cos(t) + i * \sin(t). \qquad (19.16)$$

Since our calculators cannot deal with i, we cannot check Equation 19.16 numerically. When you reach the second semester of calculus you may learn how Euler guessed it.

(a) Show that by combining Equation 19.16 with the formula:

$$e^{i(x+y)} = e^{ix} * e^{iy},$$

and remembering that $i^2 = -1$, you can figure out the addition laws for the sine and the cosine.

(b) What is the value of $e^{i\pi}$?

APPENDIX

Answers to Starred Problems

In guessing a conundrum, or in catching a flea, we do not expect the breathless victor to give us afterwards, in cold blood, a history of the mental or muscular efforts by which he achieved success; but a mathematical calculation is another thing.

– Lewis Carroll, *A Tangled Tale*

These are answers only, not model solutions whose style you can imitate. Use them just to check your work.

CHAPTER 1

3. $C = 720 + 85V$. Intercept \$720, slope 85\$/day. When $V = 14$ days, $C = \$1910$. For \$2100 you can stay 16 days.

5. (a) B: monthly bill, in \$. M: monthly use, in message units. $B = 16.50 + 0.07M$.
   (b) slope = 0.07\$/message unit, *not* \$/M.
   (e) $M = \dfrac{B - 16.50}{0.07} = 14.29B - 235.71$.

8. One-year cost: \$504.20. Daily costs reach purchase price of \$475 in 5937 days, or about $16\frac{1}{4}$ years.

12. D: dues. W: wages. $D = 85 + 0.005W$. When $W = 5000\$$, $D = 110\$$. When $D = 213\$$, $W = 25\,600\$$.

16. $T = -4 - 1.5H$. (H = hours since 5 P.M.). At 6:15, $T = -5.875°$ (or $-5.9°$). At 4 P.M. $H = -1$ and $T = -2.5°$. $T = 0°$ at $H = -2.667$ hours, or 2:20 P.M.

19. Intercept = 14 million \$. slope = 2.9 million \$/yr. 1990 prediction: sales = 72 million \$.

25. (a) $R = \dfrac{T - 4.6}{0.17} = \dfrac{1}{0.17}(T - 4.6) = 5.882(T - 4.6) = 5.882T - 27.06$.
    (b) $R = \dfrac{T + 4.6}{0.17} = \dfrac{1}{0.17}(T + 4.6) = 5.882(T + 4.6) = 5.882T + 27.06$.

(c) $R = \dfrac{T - 4.6}{-0.17} = \dfrac{-1}{0.17}(T - 4.6) = -5.882(T - 4.6) = -5.882T + 27.06.$

(d) $R = \dfrac{T + 4.6}{-0.17} = \dfrac{-1}{0.17}(T + 4.6) = -5.882(T + 4.6) = -5.882T - 27.06.$

(e) $R = \dfrac{T - I}{0.17} = \dfrac{1}{0.17}(T - I) = -5.882(T - I) = -5.882T - 27.06.$

(f) $R = \dfrac{T - 4.6}{m} = \dfrac{1}{m}(T - 4.6) = \dfrac{1}{m}\,T - \dfrac{4.6}{m}.$

(g) $R = \dfrac{T - I}{m} = \dfrac{1}{m}(T - I) = \dfrac{1}{m}\,T - \dfrac{I}{m}.$

28. $Z = 1.3(W - 4.6) + 7.5 = 1.3W + 1.52.$

31. $M = (3 - 5m)/(1 - m).$

CHAPTER 2

2. **(a)** $(12.81 - 4.67)/(2.59 - 0.63) = 4.15.$
(b) $12.81/2.29 - 4.67/0.63 = -2.47.$
(c) $(12.81 - 4.67)/2.59 - 0.63 = 2.51.$
(d) $12.81 - 4.67/(2.59 - 0.63) = 10.43.$

3. C = cost, (\$), D = number of degree days. $C = 112.50 + 0.2344D$. Intercept: \$112.50, slope: 0.2344\$/degree day. When $D = 5300$, $C = \$1354.69$.

5. $V = 35\,000 - 4000Y$, where Y is age in years. 6 years and 3 months.

7. 0.57 min/saucer. 15.4 min setup time. 1.75 saucers/min.

9. D = depth, in inches. T = time, in hours since 7 A.M. Then $D = 1.2474 + 0.6105T$. The snow falls at a rate of 0.6 inch/hour (the slope). At 10 A.M. $T = 3$ and $D = 3.08$ inches. Storm started when $D = 0$. Then $T = -2.04$, which is about 5 A.M.

12. **(a)** Salary \$66.95. Commission 8.90\$/vacuum. **(b)** \$281.

14. 8 min/page. 18 min to start up. 178 min (nearly 3 hours) to type 20 pages. 12.75 (12¾) pages in 2 hours. In 2 days I could type 358 pages–if I neither ate nor slept.

16. 1450 miles. 7500 gallons.

18. **(a)** $Q = -\frac{8}{5}P + 8.$ **(b)** $Q = -\frac{a}{b}P + a.$

24. **(a)** $Q = 3P - 20.$ $P = \frac{1}{3}(Q + 20).$

(b) $Q = \frac{1}{3}(P + 20).$ $P = 3Q - 20.$

(c) $Q = 3.02P - 22.29.$ $P = 0.33Q + 7.38.$

(d) $Q = \frac{A' - A}{5} P + \frac{14A - 9A'}{5}$. $P = \frac{5}{A' - A} Q + \frac{9A' - 14A}{A' - A}$.

(e) $Q = \frac{15}{B' - B} P + \frac{7B' - 22B}{B' - B}$. $P = \frac{B' - B}{15} Q + \frac{22B - 7B'}{15}$.

(f) $Q = \frac{A' - A}{B' - B} P + \frac{AB' - A'B}{B' - B}$. $P = \frac{B' - B}{A' - A} Q + \frac{BA' - B'A}{A' - A}$.

CHAPTER 3

3. 15-lb bag costs $3.06. 30-lb bag costs $5.66.

5. 72.5 years in year 2000. 101 years in year 2190. 200 years in year 2850. Unreasonable.

8. (a) 381 thousand dozen. 475 thousand dozen. (b) 1.62$/dozen. 1096 thousand dozen. Unreasonable.

11. (a) Not quite linear. (b) $746, $838.

13. (a) 3.23, 3.2, 3.
(b) 323 000, 320 000, 300 000.
(c) 0.00323, 0.0032, 0.003.
(d) 8.77, 8.8, 9.
(e) 877 000, 880 000, 900 000.
(f) 0.00877, 0.0088, 0.009.
(g) 53.0, 53, 50.
(h) 53.1, 53, 50.
(i) 54.0, 54, 50.

CHAPTER 4

1. 12.5 miles. 4.8 minutes. 720 minutes, or 12 hours, if I could ride nonstop. That makes no sense.

5. 1.3%. 29 000.

8. $55 100 (3 significant digits).

11. Yes, barely.

14. Rope cost is proportional to rope length. 2000$/mile.

16. 24.6 mi/gal. 0.0406 gal/mi. 1230 miles. 16.3 gal. 10.5 km/L. 0.0956 L/km.

22. $270 000 (2 significant digits).

23. 10.5 liters.

28. $W = 21L$. Units of 21 are m/cm. 85 m. $F = 4 - 0.047W$. No.

29. 12.38%/year.

33. **(a)** 7415 good chips. **(b)** 877 chips.

38. **(a)** 38 mL. **(b)** 106 mL.

43. **(a)** Length L, ft, rate R, gal/min. $R = k/L$. $k = 20$ ft * gal/min.
    (b) 1 gal/min. 7 gal/min. **(c)** 5 ft.

45. **(a)** Volume is halved. **(b)** Volume is doubled.

CHAPTER 5

1. 3.954 ~ 4.0 cents/can

3. 375 g. (The proportion is 3/4 = 0.75.)

5. $344 in high-risk, $956 in low-risk securities.

8. **(a)** 60% Sturdyturf, 40% Quickgrow. **(b)** Sturdyturf costs $1.62/lb, mixture costs $1.50/lb.

13. $C = mF + M(1 - F)$. $F = \dfrac{C - M}{m - M}$. 35%. 42%. 45% (no vermouth). ?.

15. Mixture should be 65.7% milk: add 1 gallon of water to every 2 gallons of milk!

16. The college can afford to admit 25% scholarship students.

19. $2.13/lb.

CHAPTER 6

1. Halfway through 1972 (12.4 years from 1960).

6. Adult tickets cost $3.75, childrens' $1.40.

8. $c = 2.2$ min/stop. $d = 1.3$ min/mile.

10. $N = \dfrac{D}{S - m}$.

CHAPTER 8

2. **(a)** $5 * 10^4$, $4.877 * 10^4$.
   (b) $3 * 10^{-10}$, $3.31 * 10^{-10}$.
   (c) 2.4, 2.56.

3. 36.8 nm, 0.21 μsec, 10 GW.

7. 1.3 μsec. $2.4 * 10^{13}$.

9. About 4.3.

13. Over 8000.

15. $1.7 * 10^{-24}$ g.

28. **(a)** $3.6 * 10^6$. **(b)** $3.6 * 10^{13}$. **(c)** $2.5 * 10^7$ kWh. **(d)** 29 000 years, or 29 millennia.

30. **(a)** 425. **(b)** There are a lot more. **(c)** A one-megaton bomb has the explosive power of one megaton (one million tons) of TNT. It takes 1000 one-kiloton bombs to make a megaton explosion.

CHAPTER 9

3. **(a)** 2^{A+B}. **(b)** 2^{A-B}. **(c)** 3^A. **(d)** $3^{2A+1/A}$. **(e)** $5^{(A^2-A)}$. **(f)** $5^{(A^2+1/A)}$. **(g)** 7^{2A}. **(h)** $7^{(A^2)}$. **(i)** 4. **(j)** 64. **(k)** 2431^E. **(l)** 8303^E. **(m)** 29^{6F}. **(n)** Leave it alone. **(o)** 0. **(p)** $31^{(I^J-IJ)}$. **(q)** 1. **(r)** t^{2u}.

4. Linear doubling time from start: 9 years 9 months. Exponential doubling time: 5 years 10 months.

7. **(a)** Annual growth factor = 0.8626. Price drops 13.74% per year. Calculator will cost $50 in just over 2½ years.

9. 16.8%.

11. $P = 4 * 1.02^Y$, P measured in billions of people, Y in years since 1980. In 1982, $P = 4.16$. In 1900, $P = 0.82$. In 2000, $P = 5.9$. In 3000, $P = 2.4 * 10^9$!

13. **(a)** 471 million. **(c)** Grows 1.76% per year.

16. **(a)** $63 656. **(b)** 22.42%.

18. 20% margin on original cost. 17% margin on final selling price. The numbers 60 radios and $30.00 are irrelevant.

20. 73%.

CHAPTER 10

1. Annual growth factor: 1.11. Value in 1948: $42 700.

4. Number of bacteria = $16\,500 * 1.12^H$ (H = growth time in hours). Population reaches 26 500 in about 4 hours 15 minutes.

6. **(a)** 0.95%. **(b)** Just over 6 years.

11. **(a)** 3.19%, 3.53%. **(b)** 3.31%.

15. 8.3% per year.

19. **(a)** 2. **(b)** 2.000000009 (on my calculator). Accumulated roundoff error.

CHAPTER 11

2. **(a)** 2.65. **(b)** −0.82. **(c)** −7.55. **(d)** 2.35.

5. **(a)** 3.3010. **(b)** 2.4771. **(c)** 0.6990. **(d)** 0.3495. **(e)** 0.6449. **(f)** 3.1505.

7. $A = \sqrt{\dfrac{10^5}{2}} = 223.6.$

   $B = 10^{(5/3)} = 46.4.$

   $C = B = 46.4.$

   $D = 2.$

   $E = 10^{-234.56} = 10^{0.44} * 10^{-235} = 2.75 * 10^{-235}.$

   $F = 10^{147.2} = 1.6 * 10^{147}.$

   $G = 478.70.$

   $H = 4.33.$

   $K = 0.046.$

   $L = -45.8.$

9. $K = \left(\dfrac{P}{P(0)}\right)^{1/t}, \quad t = \dfrac{\log\left(\dfrac{P}{P(0)}\right)}{\log(K)}.$

CHAPTER 12

1. **(a)** Let P = population, in millions, and measure time Y in years since 1970. Then $P = 2.3 * 1.10^Y$.
   (b) $\log P = 0.36 + 0.041Y$. **(c)** 5.6, 9.6, 40, 430.
   (d) 15.5 years from 1970, or mid-1985. **(e)** 7.3 years.

4. **(a)** $P = 100(0.992)^Y$. 86.5 g 18 years from now. 44.8 g in 100 years. 1.8 g in 500 years.
   (b) Notice that $\log(P) = -1.74$ 500 years from now.
   (c) 86.3 years. **(d)** 1400 years from now.

7. $P = 259\,000(1.053)^Y$. 26 years.

9. 25 years (2 half-lives). 125 years (10 half-lives).

12. **(a)** 10. **(b)** One million, ten million. **(c)** $0.3 * 50 = 15$ years.

14. **(a)** $\frac{1}{2}(\log(10P) + \log(P)) = \frac{1}{2}(\log(10P^2) = \log(\sqrt{10}\,P)).$
    (b), (c) $3.16 \sim 3.$

CHAPTER 13

3. 7.35 years.

8. The higher rate compounded monthly yields 13.8% annually and is better for you.

12. (a) \$10 100. (b) \$10 094.

16. Price = $73e^{-0.1Y}$. Price = \$40 in 1982.

18. Growth factors: (a) 4.06. (b) 0.73. (c) e, or 2.72. (d) 1.004.

26. $t = \dfrac{\ln\left(\dfrac{P}{P(0)}\right)}{k}$.

27. $H = 2/e = 0.736$.

CHAPTER 14

7. (a) \$100. (b) \$378. (c) 19\$/yr, 72\$/yr. (d) 190\$/yr.

9. Rate of change of price = $-9.85e^{0.1Y}$ \$/yr. −9.85\$/yr in 1976. −4.00\$/yr in 1982.

11. (a) Since bacteria live about h hours, $1/h$ of them die each hour. (To see that, think about what it says when $h = 3$.) Thus P/h of them die each hour.
    (b) $R = \quad k - \dfrac{1}{h}\, P$. (c) $P = P(0)e^{(k-1/h)t}$.
    (d) Doubling time $= \dfrac{\ln 2}{k - \dfrac{1}{h}}$.

CHAPTER 16

3. 975 cm/sec², 3622 mi/hr².

6. $D = 0.0486V^2$. Units of 0.0486: ft/(mi/hr)². 44 ft, 122 ft, 394 ft.

10. $T = 2.4 * 10^{-7}\, r^{3/2}$. Units of constant: days/mi$^{3/2}$. $r = 2.6 * 10^4\, T^{2/3}$. Units of constant: mi/day$^{2/3}$.

12. Q is multiplied by 2^P.

13. $c = 1.45$, $k = 3.01$.

CHAPTER 17

1. (a) 80 ft, −40 ft/sec. (b) 0.75 sec, 138 ft. (c) 3.6 sec, 91 ft.

4. (a) Intercept = 1272 loaves/week. Slope = −937.50 (loaves/week)/(\$/loaf).
   (b) 373 loaves/week. (c) \$0.90 per loaf. (d) \$84, \$93, \$92.50.
   (e) Profit $= L(P - 0.72) = -938P^2 + 1947P - 916$. \$1.04 per loaf.
   (f) \$0.72 per loaf, \$1.38 per loaf.

6. (a) About half a second. (b) About 4 ft.

9. 34 mi/hr, 68 mi/hr. A crash.

11. **(a)** $-1, -2$. **(b)** $1, 2$. **(c)** $-3.56, 0.56$. **(d)** $0.56, 3.56$. **(e)** Same as (c). **(f)** $-1, 3$. **(g)** no solutions **(h)** -2 (one solution) **(i)** no solutions **(j)** $(-3 \pm \sqrt{9-4C})/2$ **(k), (l)** quote the quadratic formula, Equation 17.31.

14. $G = (1 + \sqrt{5})/2 = 1.62$. $5/3 = 1.6$, so a standard file card is close to the right shape. 6.8 inches would be the correct width.

CHAPTER 18

1. 5700 feet.

3. 185 m.

6. 3.7 miles.

10. **(a)** A knot is one nautical mile per hour. A nautical mile is 6076 feet.
(b) 102 nautical miles.
(c) 11° East of North, 79° West of South.

15. $\tan(90 - A) = 1/\tan(A)$.

20. **(a)** 90°, 53.1°, 36.9°. **(b)** 82.8°, 55.8°, 41.4°.

22. 9.8 miles.

24. 213 feet, 240 feet.

CHAPTER 19

1. **(a)** 0.662. **(b)** $\pi/5$. **(c)** 0. **(d)** -2.15. **(e)** $3\pi/4$. **(f)** π. **(g)** 1.39. **(h)** $2\pi/3$. **(i)** 1.00. **(j)** $-\pi$.

2. **(a)** 86°. **(b)** 360°. **(c)** -9.918°. **(d)** 22.5°. **(e)** 0°. **(f)** 240°. **(g)** 5157°. **(h)** 30°. **(i)** 57.3°. **(j)** 184 320° (512 full circles).

3. Amplitude, period, frequency, phase:
(a) 0.67, 2π or 6.3, $1/2\pi$ or 0.16, 0.
(b) 1, 9.4, 0.11, 0.
(c) 1, 6.3, 0.16, 0.67.
(d) 1.4, 4.5, 0.22, 1.4.
(e) 1.4, 4.5, 0.22, 1.
(f) 1, 2, ½, 0.
(g) 1, 6.3, 0.16, -2.
(h) 1, 6.3, 0.16, $\pi/2$.
(i) 1, 6.3, 0.16, π.
(j) α, $2\pi/\beta$, $\beta/2\pi$, γ.

INDEX